Presented to Willie F. Mose
by
Carl Doerflinger
of Milwaukee

PLATE I.

101.

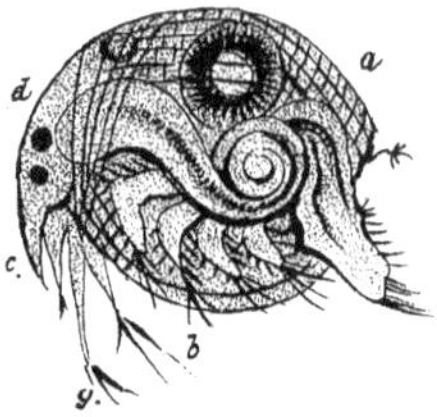

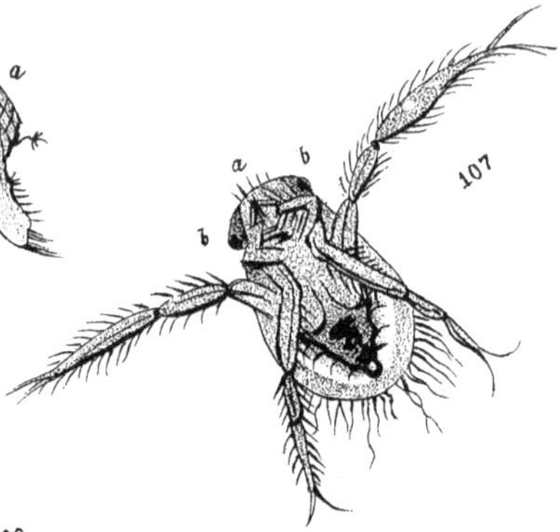

107

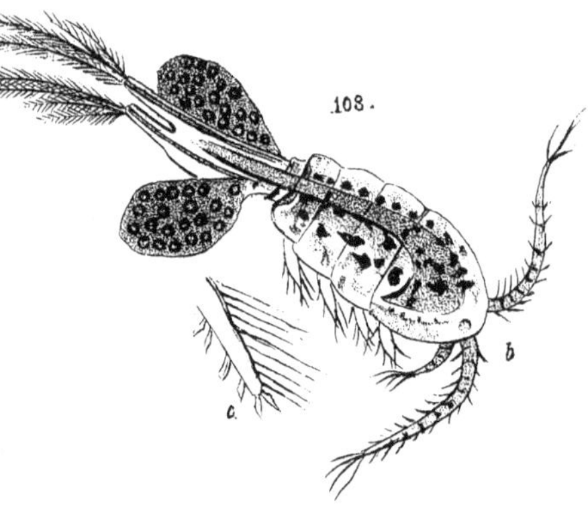

103.

102.

104

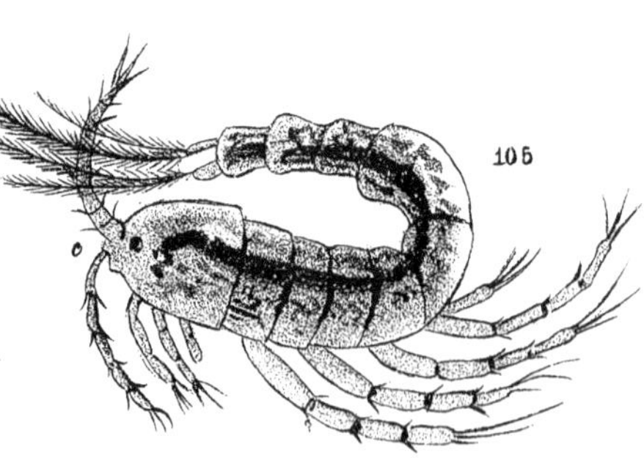

105

106.

THE

AMATEUR MICROSCOPIST:

OR,

Views of the Microscopic World.

A HANDBOOK OF

MICROSCOPIC MANIPULATION AND MICROSCOPIC OBJECTS.

BY JOHN BROCKLESBY, A.M.,

PROFESSOR OF MATHEMATICS AND NATURAL PHILOSOPHY IN TRINITY COLLEGE, HARTFORD;
AUTHOR OF "THE ELEMENTS OF METEOROLOGY," ETC., ETC.

ILLUSTRATED WITH 247 FIGURES ON WOOD AND STONE.

NEW YORK.
WILLIAM WOOD & COMPANY,
No. 27 GREAT JONES STREET.
1871.

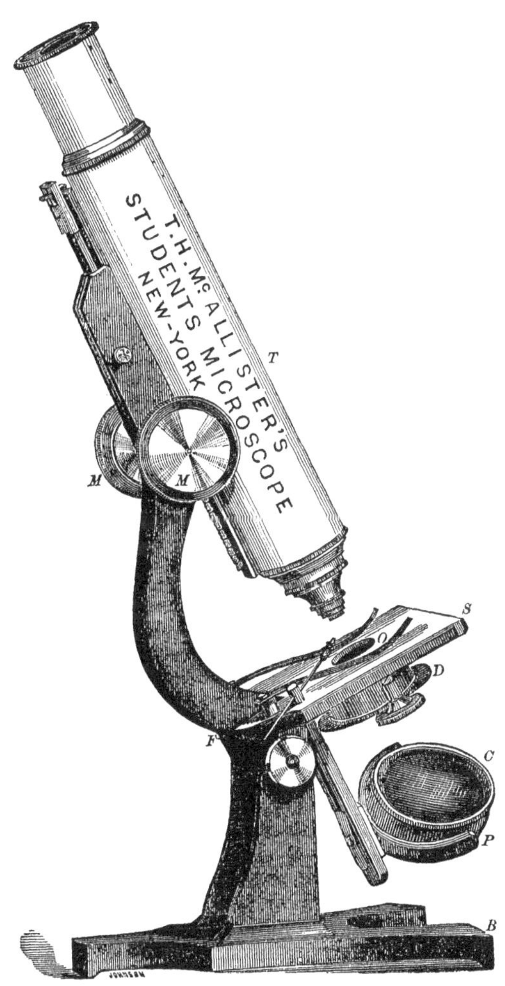
T.H. McALLISTER'S
STUDENTS MICROSCOPE
NEW-YORK
T
M
M
S
O
D
F
C
P
B

PREFACE.

I HAVE been led to believe that a popular work on the microscope and its revelations would at once be interesting and useful, and this belief has resulted in the present treatise, which simply exhibits and describes some of the most rare and curious objects of the microscopic world, and the modes of preparing them for observation under the microscope; together with a short account of this instrument.

In the preparation of this volume liberal use has been made of the discoveries of the distinguished Ehrenberg, and I have also drawn copiously from the writings of Grew, Adams, Pritchard, Mantell, Carpenter, Quekett, Hogg, Beale, and others; and from these the greater part of the illustrations have also been obtained. Without specifying other portions of the book, the chapter on crystallizations (except the remarks upon snow) is the result of my own observations, and the drawings it contains are the representations of actual crystallizations, seen and drawn by the artist. Besides these delineations many other original drawings and cuts are scattered throughout the work.

A knowledge of the wondrous revelations of the microscope cannot but be interesting; yet I trust that the perusal of this little volume may subserve a higher purpose than to while away an idle hour: that it will enkindle in the reader a desire to use this noble instrument, and by its aid to explore for himself the hidden realms of Nature. A few years ago the microscope was simply regarded as a costly toy, but now its value is appreciated in almost every department of physical science. The information it affords the physician in reference to the tissues of the human body, the nature of diseases, and the constitution of the blood, is beyond all price.

The microscope detects the "ingredients invisible to the naked eye, whether precipitated in atoms or aggregated in crystals, which adulterate our food, drink, and medicines, and reveals the lurking poison in the minute crystals which its solution precipitates."

In the department of vegetable physiology it enables the observer to study the incipient forms of vegetable life, and the structure of the most delicate tissues. To the geologist and zoologist it is indispensable, for without it they could not read the records of the rocks, and would know comparatively nothing of that luxuriant vegetation which once abounded on the globe, nor of those minute animal organisms whose remains are now entombed in the limestone strata and ranges of the earth.

Moreover, in the world revealed by the microscope we trace the workings of Infinite Benevolence, as visibly impressed on minute forms and organizations as in the starry vault emblazoned upon its rolling worlds. Here we learn with new force the harmony of Nature with Revelation, and how true it is, "that a sparrow shall not fall to the ground without our Father."

HARTFORD, CT., August 5, 1871.

CONTENTS.

CHAPTER IV.

OF THE STRUCTURE OF WOOD AND HERBS.

CHAPTER V.

CRYSTALLIZATIONS.

CHAPTER VI.

PARTS OF INSECTS AND MISCELLANEOUS OBJECTS.

VIEWS OF THE MICROSCOPIC WORLD.

INTRODUCTORY CHAPTER.

THE MICROSCOPE, AND THE MOUNTING OF MICROSCOPIC OBJECTS.

Microscopes are divided into two classes, *single* and *compound*. The *single microscope* in its simplest form is a convex lens, and the magnified image of the object passes at once to the eye of the observer; the object being magnified in the ratio of the *focal distance* of the lens to the *limit of distinct vision*, which varies from 5 to 10 inches in different persons. A single microscope is represented by Fig. 1, where A B is the lens; C D an object placed in the principal focus of the lens, at the distance H I; and E F the magnified image, seen at the distance of distinct vision by the eye at N. The image exceeds the object in length and breadth as much as N K is larger than H I.

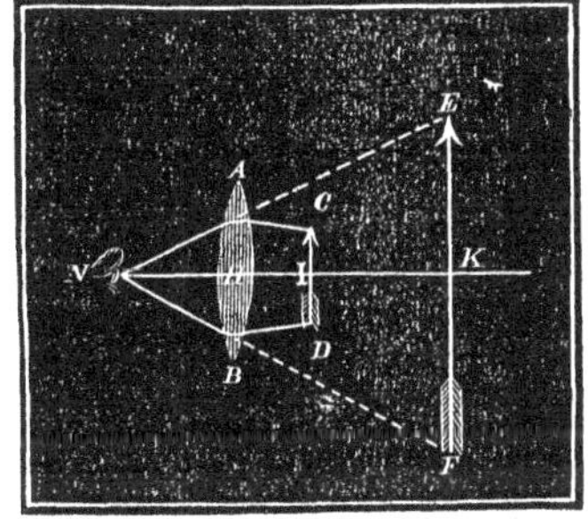

Fig. 1.

The *compound microscope* must have at least two lenses, viz., the *object-glass* and the *eye-glass*. The office of the former is to produce a magnified image of the object, which image is again magnified by the eye-glass, as if it was an object: the eye-glass being, in fact, a single microscope. The compound microscope is represented in Fig. 2, where A B is the object, D C the object-glass, and F E the image of the object formed by the object-glass, so situated as to be in the principal focus of the eye-glass G H. By this lens the divergent rays of light proceeding from the image F E, have their directions so changed, that entering the eye on the side of the lens G P H, a second magnified image is clearly discerned at K L, at the limit of distinct vision. The entire magnifying power of the instrument is equal to the combined effect of the two glasses, and is estimated as follows: The image F E is as much larger than the object A B, as its distance from the centre of the object-glass C D exceeds the distance of A B from the same

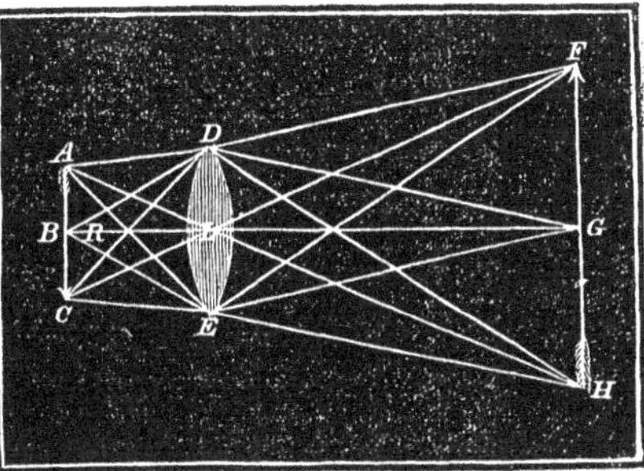

Fig. 2.

point; and the image K L is as many times greater than F E, as the limit of distinct vision exceeds the principal foca.

This is simply the rudimentary form of the compound microscope. As now constructed, both the eye-piece and the object-glass consist of a combination of lenses by which the removal of very serious optical imperfections is effected. A full description of these combinations, and of the nature of the errors they correct, can be found in any good treatise on Optics.

A very serviceable and well-constructed compound microscope, and one every way adapted to the wants of the majority of microscopists, is "McAllister's Student's Microscope," which is represented in the frontispiece.

This instrument, when inclined, as shown in the engraving, stands twelve inches high. The base B is of iron, lackered, with uprights to receive the axis upon which the body inclines. The tube T is of brass, with extension draw-tube.

The first adjustment for the focus is made by means of a delicate watch-chain controlled by large milled head-screws, M M, on each side of the tube; this adjustment is extremely sensitive and exact, even with the highest powers of the instrument. In addition to this there is a micrometer adjustment attached to the stage.

The stage, S, is made of brass, and has brass springs to hold the object. A diaphragm plate, D, is placed below the stage, and the latter is also furnished with a pair of forceps, F, which are very convenient for holding an object for examination.

The mirrors, C P, concave on one side and plane on the other, are employed for reflecting light upon the object which is placed over the opening, O, in the stage. The mirrors are so mounted that they can be turned in any direction.

ACCESSORY APPARATUS.

Micrometers.—In examining objects with the microscope it is often desirable to ascertain their exact dimensions. Measuring-instruments, called *micrometers*, are therefore sometimes applied to the object itself, and sometimes to the magnified image. When applied to the object the *stage micrometer* is employed, which consists of a slip of glass upon which parallel lines are evenly ruled with the point of a diamond, so close together that 2,500, and in some cases 50,000, occupy only the space of one inch. This apparatus is placed upon the stage, and the object examined measured by it. Of the second kind is Jackson's micrometer slide, which is so arranged in the eye-piece of the microscope that it can be brought over the magnified image by means of a screw.

The Diaphragm.—To the under surface of the stage is generally affixed a circular plate of metal pierced with holes of various diameters; this is called a *diaphragm*. It is used to modify the amount of light thrown by the mirror through the orifice in the stage upon the object. By turning the plate the holes can be brought successively under the opening in the stage.

The Illumination of Opaque Objects.—This is effected in several ways: 1st. By condensing the light upon the object by means of a large, simple, plano-convex lens, called a bull's-eye condenser. 2d. By concentrating light upon the object by means of a *metallic mirror* fitted to the side of the instrument. 3d. By causing the rays of light reflected from the mirror, and passing round the circumference of the object, to impinge upon a *concave annular reflector*, called a *lieberkuhn*. This mirror is adapted to the object-glass, and from it the rays are reflected downwards and brought to a focus upon the object itself.

THE CAMERA LUCIDA.—This instrument is fitted to the eye-piece of the microscope, and enables the observer to sketch upon a paper placed upon the table the magnified image of an object seen in the microscope. In Nachet's camera the instrument consists of a triangular glass prism, having its three faces and angles equal. Fig. 4 shows this instrument mounted in a cap which fits the top of the eye-piece.

To the accessary apparatus already mentioned may be added the animalcule cage, the machine for cutting circles of the glass, dissecting-scissors, and knives; glass-tubes, needles, the metallic stage, spirit-lamp, and hand forceps; the compressor and the machine for cutting sections of wood, and some others.

POLARIZED LIGHT.

Without entering into the subject of polarized light, it is here sufficient to say, that when certain objects are viewed under the microscope by polarized light instead of common light, they glow with the most splendid and gorgeous colors. In order that the light employed may be polarized a special apparatus is employed, which can be readily attached to the microscope. This apparatus consists of a Nichol's prism, placed below the stage of the microscope, and called a *polarizer;* and an *analyaer*, which is usually also a Nichol's prism set in a brass tube, and inserted in the body of the microscope directly behind the object-glass.

MOUNTING OF OBJECTS.

As the microscopist not merely desires to prepare objects for present examination, but also to preserve them for future inspection, various modes have been adapted to effect this end, according to the different nature of the objects to be preserved. These modes are included under the general term of "*mounting.*" *Transparent* objects are mounted upon slips of clear glass, 3 inches by 1 inch, or 3 inches by 1½ inch. There are three modes of mounting transparent objects, viz., the dry way; in some preserving fluid; and in Canada balsam. *Opaque* objects may be fixed upon a piece of black paper gummed to the slide, or mounted on flat disks of card or cork. They may also be placed in shallow cells resembling thin pill-boxes. In all cases opaque objects should be placed on a black ground.

All objects intended for microscopical observation should be protected by a cover of thin glass, in order to prevent the entrance of dust and to exclude the air. The fluid also in which many objects are placed for examination would rise in vapor, which would condense upon the object-glass and occasion great inconvenience if it were not prevented from evaporating by a thin glass cover. The glass cover should not press upon the object, lest it should impair its distinctness or destroy its structure.

This evil is avoided by placing some substance, slightly thicker than the object, around it and between the glasses, thus forming a little cavity for the specimen, which may then be covered with thin glass without risk from pressure. This cavity is called a *cell.*

Cells may be composed of various materials: for *dry* objects a small ring of paper, card-board, gutta-percha, or vulcanized India-rubber may be fixed upon the slide by cement; the object is then placed within it and covered by thin glass, which is cemented to the slide at the edges.

If the cell is to contain fluid, some substance must be employed which is not affected by moisture. Cells of this kind are formed by gold size, a solution of asphaltum dissolved in alcohol, and marine glue. In making a cell with marine glue, the glass slide must be warmed upon a metallic plate heated by a spirit-lamp. When hot enough a small piece of glue is allowed to melt upon the slide, and is moved round and round in the position in which we wish to make the wall of the cell. When the glue is allowed to cool, any excess may be removed from the slide by the aid of a sharp knife, and the glass still further cleaned by using a solution of potash. The surface of the glass to which a cement is to be applied should be roughened by grinding, as the cement adheres better than when the glass is polished. Glass is cemented together with marine glue, which is regarded as one of the best cements that can be employed. In addition to the modes already mentioned, cells may be formed from tinfoil and pieces of perforated thin glass. In making large cells of thin glass, the edges are united together by means of marine glue.

Cement.—The chief cements employed in microscopical work are *gold size*, sealing-wax varnish, a solution of shell-lac, a solution of asphaltum, marine glue, Canada balsam, gum, and French cement.

Gold size, for microscopical purposes, is made by boiling 25 parts of linseed oil with one part of red-lead, and a third part of as much umber, for three hours. The clear fluid is to be poured off and mixed with equal parts of white-lead and yellow ochre, which have been previously well pounded together. This mixture is to be added to the fluid in small successive portions and thoroughly incorporated with it; the whole is then again to be well boiled, and the clear fluid poured off for use.

Sealing-wax varnish is made by dissolving the best sealing-wax in moderately strong alcohol.

The *shell-lac* solution, which is very useful in cementing down the thin glass covers, is made by dissolving shell-lac in alcohol.

The *solution of asphaltum* is prepared by boiling together a quarter of a pound of asphaltum, and four and a quarter ounces of linseed oil, which has been previously boiled with half an ounce of litharge until quite stringy; the mass is then to be mixed with half a pint of oil of turpentine, or as much as is required to make it of a proper consistence. It is improved by being thickened with lamp-black. This cement is soluble in oil of turpentine.

Marine glue is made by dissolving *separately* equal parts of shell-lac and India-rubber in coal or mineral naphtha, and afterwards mixing the solution thoroughly by the aid of heat. It may be rendered thinner by the addition of more naphtha. Marine glue is readily dissolved in naphtha, ether, or a solution of potash.

Canada balsam is a pure turpentine, which becomes soft on the application of a gentle heat. It is chiefly employed for mounting hard dense structures; and in consequence of its great power of penetrating textures, and highly refractive properties, the structure of many substances which cannot be distinguished in the ordinary mode of examination is clearly seen when immersed in this medium. *Liquid gum* is made by placing common gum-arabic in cold water, and then keeping the bottle containing it in a warm place until the solution has become thick. This preparation is found very useful in fixing the thin glass covers, when objects are mounted in the dry way. A solution may also be made by dissolving the powdered gum in diluted acetic acid.

French cement consists of lime and India-rubber, and is very valuable for mounting large microscopical preparations. It is made in the following manner: A quantity of India-rubber scraps is carefully melted over a clear fire in a covered iron pot, and not suffered to catch fire. When the mass is quite fluid, lime in a perfectly fine powder, having been slaked by exposure to the air, is added in small quantities at a time, the mixture being kept well stirred. When moderately thick it is removed from the fire, and is then well beaten in a mortar and moulded in the hands until it has the consistence of putty. It may be colored with vermilion or any other coloring matter. The principal advantages of this cement are that it never becomes perfectly hard, thus permitting considerable alterations to take place in the fluid contained in a cell without permitting the entrance of air, and it also adheres very firmly to the glass, though the surface of the latter should be smooth.

The cements which have been described are employed for forming cells, fixing glass cells upon the glass slide, cementing the cover upon the prepared object after it has been placed in the cell, and for other microscopical purposes.

Preserving Fluids.—Objects which would lose their peculiarities by drying, can only be preserved in anything like their original condition by moistening them *in fluid;* and the choice of the fluid in each case will depend not only upon the character of the object, but also on the purpose sought in its preservation.

For the preservation of minute vegetable forms, and also of a great number of animal substances, Dr. Beale recommends the following preparation: Mix 3 drachms of *creasote* with 6 ounces of *wood naphtha,* and add in a mortar as much prepared *chalk* as may be necessary to form a thick smooth paste. *Water* must be gradually added to the extent of 64 ounces, a few lumps of *camphor* are then thrown in, and the mixture allowed to stand in a lightly covered receptacle for two or three weeks, with occasional stirring; after which it should be filtered, and preserved in well-stopped vessels for use.

Of late years *glycerine* has been much used as a preservative fluid, as it allows the colors of vegetable objects to be restored. The best preparation of the kind is made by mixing one part of *glycerine* to two parts of *camphor-water.* *Deane's Gelatine* is one of the most convenient fluids for preserving the larger forms of confervæ and other microscopic algæ. This is prepared by soaking 1 ounce of *gelatine* in 4 ounces of *water,* until the gelatine is quite soft, and then adding 5 ounces of *honey,* previously raised to boiling heat in another vessel; the whole is then to be made boiling hot, and when it is somewhat cooled, but is still perfectly fluid, 6 drops of *creasote* and ½ an ounce of *alcohol,* previously mixed together, are to be added, and the whole is then to be filtered through fine flannel. When required for use the mixture must be slightly warmed, and a drop placed upon the preparation on the glass slide, which should also be warmed a little. Next, the glass cover, having been breathed upon, should be carefully laid on, and the edges covered with a coating of asphaltum cement.

A mixture of *gelatine* and *glycerine,* in equal parts, has been used for the same purpose. A solution of *chloride of calcium* in three parts of *water,* is also employed by vegetable microscopists. This last solution has the disadvantage of not preserving color, but the advantage that it does not dry up in the cell.

For the preservation of *animal* tissues, a mixture of one part of *alcohol* to five parts of *water* answers tolerably well.

The best fluid for this purpose is, in most cases, Goadby's solution, which is made by dissolving 4 ounces of *coarse sea-salt*, 2 ounces of *alum*, and 4 grains of *corrosive sublimate* in 4 pints of *boiling water*. This mixture should be carefully filtered before it is used, and, for all delicate preparations, it may be diluted with an equal volume, or even with twice its volume of water. This solution must not be used where any calcareous texture, as shell or bone, forms any part of the preparation. It is very valuable for preserving anatomical specimens.

A solution of chromic acid is well adapted for preserving many microscopical objects. It is particularly useful for hardening portions of the nervous system previous to cutting them into thin sections for examination. This solution is prepared by dissolving a sufficient amount of the crystals of the acid in distilled water to make the liquid of a pale straw-color. "It is often quite impossible," says Dr. Carpenter, "to tell beforehand what preservative fluid will best answer for a particular kind of preparation, and it is always desirable, where there is no lack of material, to mount the same object in two or three different ways, marking on each slide the method employed, and comparing the specimens from time to time, so as to judge how each is affected."

Having alluded to the different ways of mounting objects, explained the mode of preparing cells, and described the various cements and preservative fluids used for microscopic purposes, we will now unfold more fully to what transparent objects the different modes of mounting are adapted, and also the process of mounting.

Mounting in the Dry Way.—There are certain objects which, viewed by transmitted light, are more advantageously seen when simply laid on glass than when immersed in fluid and balsam. This is the case with sections of bones and teeth, and with the scales of Lepidopterous and other insects whose surface-markings are far more distinct than when mounted in any other way. These objects, however, must be protected by a glass cover, so attached to the glass slide as to keep the object in its place, besides being itself secure. To effect the latter object, gold size mixed with lamp-black can be employed. If the object has any tendency to curl up, and to keep off the cover by so doing, it will be useful, while applying the gold size, to press the two pieces of glass together by means of a spring clasp, such as is used for fastening clothes to the line when drying. The slide should be kept clamped until the gold size is sufficiently dry to hold down the cover by itself.

Another mode is to put a little dot of ink in the very centre of the slide, on the side opposite to the object, and directly below the place it is intended to occupy, so as to act as a guide during the process of mounting. Now place the object carefully over the ink-dot, put the thin glass cover very lightly upon it, and fasten it down with two or more strips of thin paper, which are to be pasted or gummed round its edges. Two dry objects may be mounted upon the same slide.

Mounting in Fluid.—As a general rule, it is desirable that objects which are to be mounted in fluid should be soaked in the particular fluid to be employed for some little time before mounting; since, if this precaution be not taken, air-bubbles are very apt to appear.

The tissues of animals, parts of insects, and vessels of plants, and all such objects as would lose their characteristics if prepared in any other way, are mounted in fluids. The process is as follows:

Clear the glass slide with a weak solution of ammonia or potash, and wipe it dry with a piece of chamois leather or cotton velvet. Put a drop of the preserving fluid into the cell, then place the object in it with a small pair of forceps, and spread it out very carefully with the point of a needle, taking care to remove any air-bubbles that may appear.

The thin cover may now be placed upon the cell, one side being first brought down upon its edge, and then the other, and if the cell has been previously running over with the fluid, it is not likely that any air-space will remain. All superfluous fluid is then to be taken up by blotting-paper, and the surface of the cell and the edges of the cover made perfectly dry. When this is done, the cell may be closed by cementing the edges of the cover and the sides of the cell by a thin layer of size or asphaltum.

Mounting in Canada Balsam.—Among the various objects which are advantageously mounted in this way are sections of shell, dry vegetable preparations, the hard portions of the structure of insects, the horny tissues of the higher animals, and also many organized substances, both recent and fossil. The great obstacle encountered in mounting objects in Canada balsam is to keep them free from air-bubbles, but by proceeding in the following manner this difficulty is usually overcome: Take a curved glass tube, put some balsam into it, cork up the wide end, and let it stand upon its end until wanted. Upon the metallic table place the glass slide, light the spirit-lamp and put it under the table, so as to warm the slide throughout, but not to overheat it. Then take the object, as a seed for instance, put it into some spirits of turpentine, and let it wait there while the slide is becoming warm. The next process is to remove the lamp and to hold the tube containing the balsam over the flame, when the balsam will immediately run towards the orifice at the small end, and a drop will ooze out. This drop should be placed on the centre of the slide, and the seed taken out of the turpentine is then laid upon the warm balsam and gently pressed into it with a needle. With the needle turn the seed about so as to make sure that no air-bubbles are clinging to its surface. Now take a thin glass cover, perfectly clean, warm it over the spirit-lamp, and lay it carefully and slowly upon the balsam. Press it slightly down, and then, having placed on the cover a small circle of pasteboard, put the slide within the clothes'-line clamp already mentioned, and then lay away the slide till the balsam has hardened. The superfluous balsam which will be found adhering to the edges of the glass cover may now be removed by scraping it away with a knife, and then the slide more thoroughly cleaned by wiping it with a rag moistened with spirits of turpentine or ether.

Of Injections.—The arrangement of the minute vessels or capillaries distributed throughout various textures is not to be demonstrated, in all instances, by the usual method of investigation, in consequence of the transparency of the walls of the tube, and can only be well shown by means of injections of coloring matter. The art of making these preparations is one in which success can only be attained by long practice; and better specimens can therefore be obtained from those who have made it a business than are likely to be prepared by amateurs. Injections form very interesting objects for microscopic examination.

Any one who wishes to study more in detail and in all their minutiæ the various subjects which have just been discussed, can consult with great advantage such works on microscopy as those of Quekett, Carpenter, Hogg, Beale, and others.

Figure 3.

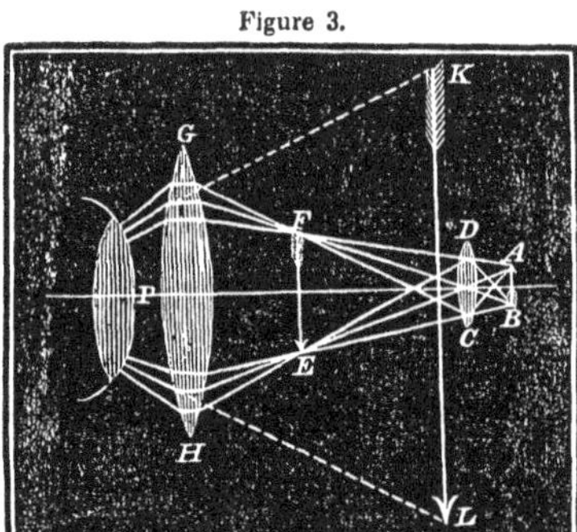

Figure 4.

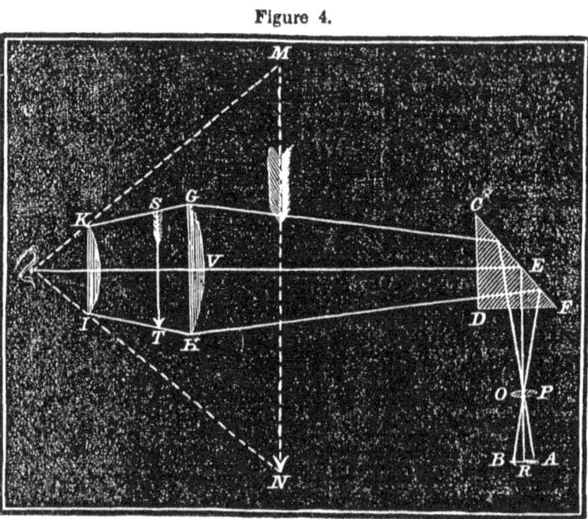

CHAPTER I.

INFUSORIAL ANIMALCULES AND PROTOPHYTES.

"Full Nature swarms with life; one wondrous mass
Of animals, or atoms organized,
Waiting the vital breath when Parent-Heaven
Shall bid His spirit blow. The hoary fen,
In putrid streams, emits the living cloud
Of pestilence. Through subterranean cells,
Where searching sunbeams scarce can find a way,
Earth animated heaves—and where the pool
Stands mantled o'er with green, invisible,
Amid the floating verdure millions stray.
Each liquid too, whether it pierces, soothes,
Inflames, refreshes or exalts the taste,
With various forms abounds. Nor is the stream
Of purest crystal, nor the lucid air,
Though one transparent vacancy it seems,
Void of their unseen people."—*Thomson.*

THE name of *infusorial animalcules* has been given to various species of minute living beings, which were first discovered in vegetable infusions; that is, in water containing vegetable matter. From the latter circumstance they received the appellation *infusorial*, and in consequence of their being exceedingly small they were termed *animalcules* or *little animals.*

It was supposed by the earlier naturalists that the animalcules, whose existence was thus detected, were confined to certain infusions; but it is now well ascertained that there is no necessary connection between them and the vegetable ingredients, except to this extent: that the latter, under favorable circumstances, may perhaps facilitate the development of the eggs of these living atoms, and afford a proper nourishment for the animalcule, through all the stages of its existence.

As this department of nature became more thoroughly and widely explored, new species of animalcules were discovered, and the general term of infusorial animalcules has therefore been so enlarged in its signification as to embrace, not only that class of minute beings that are found in vegetable infusions, but all those that possess the same marked peculiarities of structure, wherever discovered. In the gushing fount, the rippling brook, and the placid waters of the lake, infusorial animalcules exist in countless numbers—often swarming to such an extent as even to color the element in which they live. One species tinges the water with a blood-red hue, another causes it to appear of an intensely vivid green; while a bright yellow hue indicates the presence of a different species. They are likewise found in strong acids, and in the fluids contained in animal bodies and living plants, and have also been detected alive in moist earth, sixty feet below the surface of the soil. The broad rivers are their home, and far from shore, upon the tropic seas, the ocean swarms, for leagues, with their congregated myriads; and as the bark of the mariner nightly cuts the wave, the dazzling track it leaves upon the waters, and the fiery spray that flashes from its bows, tell of the presence of life enshrined within an infinity of living atoms. Nor is the bed of the ocean without its minute inhabitants; for the mud brought up by the deep sea lead, from the depth of six teen hundred feet, is full of organic life. There is also every reason for believing that the atmosphere abounds with the eggs of animalcules, as it does with the seeds of minute plants; and that these germs, being inconceivably light, are raised by evaporation, and borne about by the winds in unseen clouds; ready to burst into life whenever a concurrence of favorable circumstances facilitates their development. Lifted at one time to the loftiest mountain tops, at another carried down to the lowest dells and deepest caverns; they cross seas, sweep over continents, and interchange climes and seasons. In this manner are these invisible forms disseminated over every part of the world; for wherever investigations have been prosecuted, infusorial animalcules have been discovered.

Through the patient and presevering labors of distinguished naturalists, no less than as many as several hundred different species of animalcules have been distinctly recognized and delineated, and grouped into families and classes; distinguished from each other by their forms, manner of progression, habits, and modes of reproduction. One kind, the Ophrydinæ, is found embedded in vast numbers within a gelatinous mass of matter of a greenish hue, which is sometimes adherent and sometimes free, and may attain a diameter of four or five inches, presenting a strong resemblance to frogs' spawn. These masses result from the individual animalcule propagating by self-division; each living atom being connected with the rest by a gelatinous exudation from the surface of its body. Other infusoria possess the power of changing their forms at will, and in the space of a few minutes pass through a variety of curious and grotesque shapes.

Another class shoot up in the form of beautiful shrubs, crowned with bell-shaped flowers, whose margins are encircled with a fringe of slender hairs; but the flower-cups are living beings, and the mimic tree is instinct with vitality in every branch. At one moment, it is seen spreading outward and upward from the base, with all its living flowers in full expansion; and at the next, should danger threaten, every shoot suddenly contracts, and the whole group of animalcules shrink down in spiral coils, into the smallest compass. The great variety of form possessed by these interesting objects can only be fully conceived by examining those works in which they are accurately delineated. In the great work of Dr. Ehrenberg, they are beheld in all their beautiful and singular proportions. This splendid volume, of folio size, contains sixty-four plates, filled with several hundred infusorial shapes, drawn and colored from nature; all of which he regarded as true animals. Some resemble globes, trumpets, stars, boats, and coins; others assume the forms of eels and serpents; and many appear in the shape of fruits, necklaces, pitchers, wheels, flasks, cups, funnels, and fans.

But the minuteness of these beings is no less surprising than the diversity of their forms. Myriads of the Epistylis botritis, which belongs to the family of the flower-cup animalcules, might be contained in a drop of water, for it has been computed that within this small space, not less than *thirteen millions* could be comprised; and this calculation is not to be regarded as unworthy of confidence, inasmuch as the Epistylis is known to attain a length no greater than the *twenty-four hundredth part of an inch.* In the case of the Ophrydinæ already described, a cubic inch of the gelatinous matter in which they are embedded has been estimated to contain no less than eight million distinct beings; a fact which, when taken in connection with others of the same nature, render it highly probable that the living beings of the microscopic world surpass in number those which are visible to the naked eye.

STRUCTURE.—The outer covering of infusorial animalcules is of *two* kinds; the first soft and yielding, resembling the skin of the leech and slug, and so far capable of expansion and contraction as to adapt itself to the state of the animalcule whether distended or not; the second presenting the appearance of a firm, transparent shell; yet possessing a flexibility like horn. Those animalcules that are protected by the latter integument are termed *loricated*, from the Latin word *lorica*, a shell; while the name *illoricated* or *shelless* is assigned to those which are invested with the softer and more perishable covering. The materials that compose the shell vary in different species; in many instances it consists entirely of flint, and in others of lime united with oxide of iron; in some cases it is combustible and in others not. In several kinds, the lorica, in the form of a jar or cylinder, entirely surrounds the animalcule, while in others it is shaped like a shield, and protects the living atom to which it belongs, as the shell of the turtle defends its sluggish inhabitant from external danger.

Many animalcules which may be classed as illoricated have yet a firmer covering than others belonging to this division. It consists of a gelatinous membrane having a bell-shaped cylindrical or conical figure, closed at the lower end, but open at the other, through which opening the animalcule may protrude itself.

It was formerly believed that the smaller species of animalcules were entirely destitute of external organs; but such improvements have now been made in the construction of microscopes, and the organization of the living objects has been rendered so much more distinct, from the practice of feeding them on colored substances before examination, that this supposition has been shown to be entirely unfounded.

These external organs vary in kind in different animalcules, but the one which is the most remarkable, and is common to all Infusoria, is a slender filament like a hair, situated near the mouth, and from its striking resemblance to an eyelash is known by the name of *cilium*, the Latin word for eyelash.

The cilium is employed by the animalcule for the purpose of motion, and also for that of procuring food. Using this member as an oar, the creature moves swiftly through the water, and so curious is the action of this propeller, that the very stroke which effects a progressive motion, causes at the same time a current to set towards the mouth of the animalcule, bearing its prey and food within its reach.

In addition to the offices of the cilia* just described, they are supposed by some naturalists to be the principal instruments for respiration to the Infusorial world; inasmuch as similar appendages are found encircling the gills or beard of the oyster and muscle, and other animals of the like nature. It is by means of the gills that these creatures inhale the air contained in the water, and the cilia, by causing currents to

* *Cilia*, the plural of *cilium*.

flow towards these organs, furnish a continual supply of fresh air. According to Mantell, "recent discoveries have shown that cilia exist also in the internal organs of man and other vertebrated animals, and are agents by which many of the most important functions of the animal economy are performed."

When an animalcule is examined, this delicate member easily eludes observation, but if the creature is placed in a drop of water colored with indigo or carmine, the little whirls and currents created by the action of the cilia are readily detected under the microscope; and upon the evaporation of the water from the glass slide, a fine streak upon the surface indicates its existence and position.

These slender organs are variously arranged in different species of Infusoria. In some they are extended in rows throughout the entire length of the animalcule, and in others are distributed over the whole surface of the body. Fringes of cilia encircle the mouths of some, while in many kinds, the circles of cilia forming into bands, surround certain projections issuing from the upper part of the body. Numerous species are furnished with only two of these filaments projecting from the mouth, and nearly equal to the body in length. The base of each cilium terminates in a bulb, and when the organ is in motion its point describes a circle, while the globular base simply rolls round upon the surface to which it is attached. An idea may be gained of this motion by holding the arm out stiffly and swinging it round, so as to describe a circle in the air with the point of a finger; the arm then corresponds to one of the cilia, and the ball of the shoulder-joint to the bulb upon which the cilium turns. The motion is doubtless performed by muscles, and Ehrenberg considers that he has not only discovered their existence in some of the larger Infusoria, but also the arrangement of the fibres that compose them.

The bands and coronets of cilia which encircle certain classes of animalcules present, when in motion, a singular appearance. Though each organ is stationary and revolves only around its bulb, yet the combined action of the circular rows is such that they appear to revolve together, like a wheel upon its axle, and so complete is the illusion that the name of wheel-animalcules, or Rotatoria, is given to those which possess this peculiarity.

Besides these organs, stiff hairs or bristles are found upon animalcules, which, unlike the cilia, are devoid of rotation, but serve as supports to the body, and also aid these living atoms in climbing. Animalcules are also found with hook-like projections extending from the under side of the body, which are capable of motion to some extent, but do not possess the peculiar movement of the cilia. Many Infusoria are also endowed with another kind of member, that more completely subserves the purpose of motion, and which they have the power of protruding or withdrawing at pleasure, as the snail extends and retracts its horns. These organs are soft, and by some species can be thrust out from every part of the body; while in others, that are partially covered by a shell, they are confined to the uncovered portions.

The power of extension possessed by Infusoria over these organs is much greater, in proportion to their size, than in the case of snails and animals of a similar nature.

In those Infusoria that are gifted with the highest organization, as the wheel-bearing animalcules, there appears to be a member resembling a claw, by means of which they attach themselves firmly to any object within their grasp. The claw is appended to an extended portion of the body, resembling a foot.

Classification.—Dr. Ehrenberg divides this living world into two great classes, distinguished from each other by their structure: viz., the Polygastrica* or *many stomached animalcule*, and the Rotatoria† or *wheel-animalcule.* Other microscopists, however, prefer to substitute the term Infusorial animalcules for the Polygastrica.

POLYGASTRICA.—If an animalcule of this class is viewed by the microscope, a number of round spots within its body will be readily detected, which are often quite large compared with the size of the living atom. These spots are so many stomachs, connected together by a single tube, and forming the digestive apparatus of the creature. If the water around the animalcule is clear, the stomachs will appear more transparent than the rest of the body; but if it is tinted with sap-green or carmine (which substances are usually employed) they will be seen more distinctly; for the animalcule readily imbibes the colored fluid, and the stomachs from their transparency then appear of the same hue as the liquid; while the tint of the more solid portions of the body remains unchanged. The number of stomachs varies in different species.

In the annexed cut a highly magnified view of a *bell-shaped animalcule* is presented, in which the stomachs and coronets of cilia are distinctly exhibited.

None of this class of infusoria are more than the twelfth of an inch long, and the smallest species, when full grown, do not exceed in extent the *thirty-six thousandth* part of an inch. Uniting, however, in infinite multitudes, the more minute kinds form various colored masses, several feet in length. The young of many species are doubtless too minute to be visible even under the highest powers of the microscope.

Figure 6.

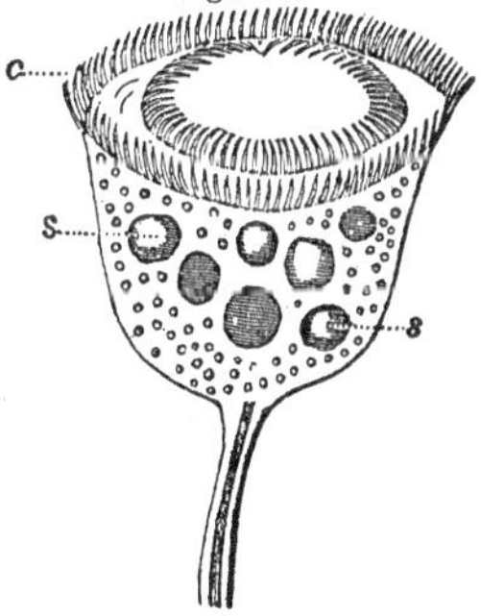

A BELL-SHAPED ANIMALCULE.
C. Cilia. S. The Stomachs.

According to Ehrenberg the Polygastrica form a very extensive division of the Infusoria; but the investigations of later microscopists have shown that a large number of classes of organisms, which Ehrenberg regarded as true animals, are simply vegetable structures, as will be more fully explained in the following pages. The number of species of the true Polygastrica is therefore much less than it was formerly supposed to be, and, consequently, the forms and conditions of life at first assigned to them are now only partially true.

Some of the Polygastrica are loricated, and others illoricated. The shells are differently formed. In one kind it consists of small rings between which cilia are situated, and in another the lorica has the form of a shield.

ROTATORIA.—The second class of Infusoria have received the appellation of Rotatoria, as has already been stated, from the circumstance that the circles of cilia which surround the upper part of the body of the animal appear when in motion to revolve like a wheel. The cilia are found upon no other portion of their body, while in the Polygastrica they are distributed over the entire surface. In some species the crowns of cilia consist of a single set, and in others several

* From the Greek polus, *many*, and gastēr, *a stomach.* † From the Latin rota, *a wheel.*

circular rows of different forms are distinctly noticed. This class of Infusoria is endowed with a highly perfected organization, and on account of their comparatively large size, some of them attaining a length of one-thirtieth of an inch, both their external and internal structure are well revealed by the microscope. The Rotatoria possess a single stomach, and many kinds are furnished with jaws and teeth, which, together with other parts, will be particularly described hereafter, when treating of individual animalcules belonging to this class. The preceding cut, Fig. 7, is, however, given at present for the sake of

Figure 7.

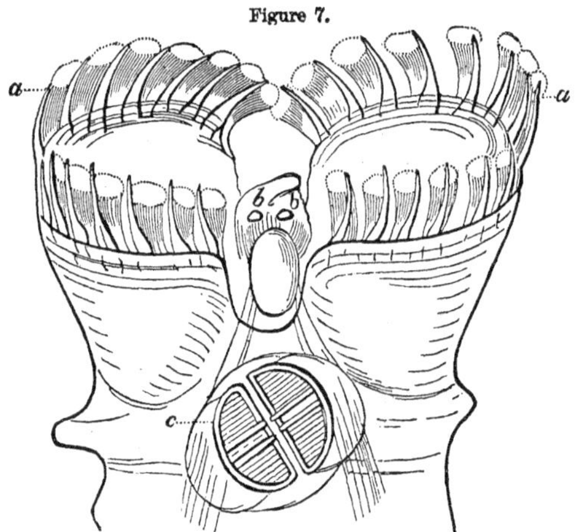

aa. The Cilia. bb. The Eyes. c. The Jaws and Teeth.

illustration. It displays the upper part of a common wheel animalcule, with the circles of cilia, jaws, teeth, and eyes, all highly magnified.

The Rotatoria reside chiefly in water, but are frequently found in moist earth, and some species have been detected dwelling in the cells of mosses and sea-weed.

EYES.—By the aid of the microscope, as now perfected, naturalists have discovered *eyes* throughout the entire class of Rotatorial animalcules; these organs, minute as they are, being well defined,—a fact that indicates the existence of a nervous system in these living atoms. The eye of larger animals is known to be an organ exceedingly complicated in its structure, and replete with the most beautiful contrivances to insure perfect vision. There is every reason for believing that the same is true of the eyes of animalcules; and if this is so, how can we sufficiently admire the wondrous perfection and consummate skill which the Creator has deigned to bestow upon some of the least of his revealed works; and above all, the unwearied benevolence displayed in every manifestation of His infinite power! In respect to the Polygastrica, Ehrenberg claims that the red-colored specks, which he observed in many organisms that he included in this class, are true eyes; but from this conclusion other microscopists dissent. Ehrenberg insists much on the red color of the speck as a distinctive indication of a visual organ, but other colors prevail in the known eyes of insects. If it is admitted that these eye-specks have a general sensation of light, their optical nature is not proved, since organisms destitute of these exhibit a like sensibility of the presence of light. The eyes of the Rotatoria are generally red, and most of the animalcules belonging to this class have two of these organs. In some three are perceived; while one kind especially has the benefit of seven or eight on each side of the head. A diversity exists in the arrangement of the eyes; in many instances they are placed in a line, side by side, in others they form a triangle. In some animalcules they are arranged in a circle, and in two species they unite in clusters on each side of the head.

Reproduction.—Animalcules multiply in several ways. First, they proceed from eggs. Secondly, they are brought forth alive. Thirdly, they increase by the growth of buds issuing from the body of the parent; the buds sprouting out and becoming themselves perfectly organized animalcules. Lastly, they are propagated by self-division, the body of an animalcule separating into two or more individual beings.

The same animalcule is not always confined to a single mode of reproduction, but its countless offspring may come into existence by one or more of the ways just detailed; one part being produced by self-division, another from eggs, and the remainder originating in buds. This circumstance accounts for the amazing fecundity of the Infusoria, which almost surpasses belief; even when limited to one method of increase. The Rotatorial animalcules are propagated from eggs alone, and but a few of these are deposited at a time; yet so quickly are the young matured, that from a single animalcule millions will proceed in the course of a few days. Dr. Ehrenberg kept one of this class of Infusoria for eighteen days in a separate vessel of water; during this interval it laid four eggs a-day, and the young, when two days old, laid the same number; so that, continuing to multiply at this rate, it was found that, under favorable circumstances, the offspring of a single creature would amount, in the course of ten days, to one million; in eleven, to four millions; and in twelve days, to sixteen millions. This is the greatest increase that has actually been determined by experiment. In the other class of Infusoria, the Polygastrica, the fecundity is much greater, inasmuch as they are endowed with more varied powers of reproduction. A single animalcule belonging to one of the larger kinds is known to separate into eight, in the course of a day, by self-division; so that, multiplying at this rate for the space of ten days, it would increase to more than one hundred and fifty million individuals. It likewise propagates from eggs, which are seen in clusters like the spawn of fishes; and also by buds, that sprout from the sides of its body. When all these facts are taken into view, it is evident that the vast number of living beings, which in a few days spring from a creature as prolific as this, is utterly beyond our powers of appreciation. In truth, this animalcule, which is called the Paramecium, increases so rapidly in stagnant waters, in the ways that have been mentioned, that some naturalists have supposed that they were produced spontaneously from inert matter.

Life and Resuscitation.—Infusorial animalcules live but a short period, for, although the duration of their life varies in different kinds, it extends only from a few hours to several weeks. Wheel animalcules have been seen in the enjoyment of their existence twenty-three days after their birth. The death of Infusoria is usually sudden, and in the larger species is attended with spasms. The soft parts rapidly decompose after death, and all the curious and elaborate organs of these singular beings entirely vanish; nothing appearing to remain except the firm, enduring shells in which many kinds of the Infusoria are encased. But, in numerous instances, this death is but apparent; the decay of the body does not take place, and in the minute speck that lies before us like an atom of inanimate dust the mysterious principle of life is still in existence. The creature may remain motionless for months, and even years; but when it is again subjected to influences favorable to its resuscitation, it awakens from its torpor, and life, with all its former energies, is once more fully displayed.

This surprising phenomenon is supported by undoubted proof. When the water containing a wheel-animalcule evaporates, the creature apparently expires, becomes dry and hard, and may be preserved in this state for years, if buried in sand. When placed in water in this condition, it will revive in a few minutes, and soon swim about with its wonted activity. In 1701 Leuwenhoeck observed this fact in wheel animalcules, and revived some specimens after keeping them dry for twenty-one months. Baker obtained the same result after a longer time, and Prof. Owen was present at the resuscitation of an animalcule after it had lain dormant in dry sand for *four years.* Nor is this all; for such is the tenacity of life in these minute beings, that the same animalcule may repeatedly pass through these phases of existence before it really expires. Mantell remarks that some wheel animals were alternately dried and rendered torpid, and then again revived *twelve times*, and at each resuscitation were as active as at first.

The eleventh revival was witnessed by Spallanzani; and he leads us to infer that, upon moistening a portion of sand containing wheel-animalcules for the *fifteeenth* time, many of them once more awoke from their stupor; but this was the last effort of vitality, for upon being dried and moistened again, no resuscitation occurred. The wonderful legend of the Seven Sleepers is here more than realized; and in the Infusorial world the romantic fiction of Rip Van Winkle becomes a sober statement of fact. Thus it is that Fancy in her wildest flights seldom sweeps beyond the circle of truth.

It is the opinion of Dr. Ehrenberg, in regard to this subject, that if the animalcule is entirely dried up and its natural heat lost, life is extinguished, but if this is not the case the creature will remain in a torpid and motionless state, capable of being revived; its body wasting away to an extent equal to the amount of nourishment necessary for the support of its life.

Influence of Temperature.—Infusorial animalcules are capable of existing throughout a great range of temperature, but eventually perish under extreme degrees of heat and cold. If water, filled with Polygastric Infusioria, be gradually raised to a temperature of 125° Fah., these creatures will live. If the increase of temperature be sudden they die at 140° Fah., notwithstanding it be kept up only for half a minute. The Rotatoria, when put in boiling water (212° Fah.), are killed, but they retain their power of revival when the water has a temperature varying from 113° to 118° Fah. Some kinds, however, are extremely sensitive, and are unable to endure an ordinary degree of warmth. This is the case with the Bell-flower animalcule, which dies under examination in a hot room. Most of the Polygastrica retain their vitality at temperatures considerably below the freezing point; but when the mercury descends as far as 7° or 8° Fah., many species can no longer exist. One kind of the Bell-flower animalcule still lives after being exposed to a temperature of 8° Fah. and the ice then gradually thawed in which it was frozen; but not more than one individual in a hundred can survive this ordeal. The Rotatorial animalcules are more susceptible, and perish when the cold is less severe.

During the Antartic expedition under Capt. James Ross, animalcules were found existing in great abundance in those inclement regions. In the sediment obtained from melted ice, floating in round masses, in the latitude of 70°, more than fifty species of loricated Infusoria were discovered alive, notwithstanding the extreme cold to which they had been exposed. According to Dr. Ehrenberg, when a layer of clear ice containing animalcules is examined under a low temperature by the microscope, each animalcule or group will be seen surrounded by a very small portion of water, which he supposes is prevented from freezing by the natural heat of their bodies; and he likewise believes that death inevitably ensues whenever the cold is sufficiently intense to congeal this enclosing film of water.

Air.—Air is as necessary to existence of Infusoria as to any other class of animated nature; for when they are denied access to the atmosphere, and are thus prevented from receiving constant supplies of pure air, life becomes extinct within a short time. If oil is poured upon the water containing animalcules, and the surface of the fluid is entirely covered with the oil, the air is necessarily excluded and the creatures speedily die. Or should the naturalist fill a phial with water in which animalcules reside, and leave it corked tightly for any length of time, he will have the mortification of finding, on examination, that the fluid, once so full of life and activity, has become entirely inert, and that millions of existences have passed away. The fact that air is necessary to the existence of Infusoria has been particularly noticed in regard to the larger kinds of wheel-animals; for when experiments have been made by placing these creatures under the receiver of an air pump, they have always ceased to live soon after the air has been withdrawn from the vessel in which they were contained.

Dr. Ehrenberg affirms, that if animalcules are placed in nitrogen gas they exist for a longer time than if they are immersed in carbonic acid or hydrogen gas. In the fumes of sulphur they quickly perish.

Poisons.—The most powerful poisons, which mingle simply in a *mechanical* manner with water, like earth, do not affect the lives of animalcules placed in the mixture; but those which unite *chemically*, and are dissolved in the fluid, soon deprive them of their existence. One kind of Infusoria has been known to live so long in water with which calomel and corrosive sublimate had been mixed, that it was doubtful whether their death was to be attributed to the effect of those ingredients or not.

Many species of animalcules can adapt themselves to a gradual change in the nature of the element in which they live, but a sudden transition kills them. For instance, similar kinds are found at the heads of rivers where the water is fresh, and their mouths, where the streams mingle with the briny ocean.

If sea-water, abounding with marine animalcules, is mixed by *slow degrees* with the fluid in which fresh-water species reside, the latter survive; but if the mixture is *suddenly* made, they perish immediately.

PHOSPHORESCENCE OF THE SEA.—Various opinions have at times been entertained in respect to the cause of this beautiful phenomena; but it is now correctly attributed chiefly to the presence of animalcules, which crowd the waters in vast multitudes. This appearance, although confined to no particular part of the ocean, attains its greatest splendor in the tropical climes, where the spectacle is often exceedingly grand and beautiful.

This brilliant phenomenon is thus graphically described by Darwin in his "Voyage of a Naturalist:"—"While sailing a little south of the River La Plata, on one very dark night, the sea presented a wonderful and most beautiful spectacle. There was a fresh breeze, and every part of the surface, which, during the day, is seen as foam, now glowed with a pale light. The vessel drove before her bows two billows of liquid phosphorus, and in her wake she was followed by a milky train. As far as the eye reached, the crest of every wave was bright, and the sky above the horizon, from the reflected glare of these livid flames, was not so utterly obscure as over the vault of the heavens. Near the mouth of the Plata some circular and oval patches, from two to four yards in diameter, shone with a steady but pale light, while the surrounding water only gave out a few sparks. The appearance resembled the reflection of the moon, or some luminous body, for the edges were sinuous from the undulations of the surface. The ship, which drew thirteen feet of water, passed over without disturbing these patches; we must therefore suppose that some of the luminous marine animals were congregated together at a greater depth than the bottom of the vessel, or thirteen feet beneath the surface of the sea."

The same phenomenon is thus depicted in the glowing language of Colton:—"We had last night a splendid exhibition of aquatic fire-works. The night was perfectly dark, and the sea smooth, and you might see a thousand living rockets shooting in all directions from our ship, and running through countless configurations, return to her, leaving their track still bright with unextinguishable flame. Then they would start again, whirling through every possible gyration, till the whole ocean around seemed medallioned with fire. We had run into an immense shoal of porpoises and small fish; the sea being filled at the same time with animalculæ, which emit a bright phosphoric light when the water is agitated. The chase of the porpoises after these small fish created the beautiful phenomenon described. The light was so strong that you could see the fish with the utmost distinctness. They lit their own path like a sky-rocket in a dark night; and our ship left the track of its keel in the wave for half a mile. I have witnessed the illumination of St. Peter's, and the castle of St. Angelo, at Rome, and heard the shout of the vast multitudes as the splendors broke over the dark cope of night, but no pyrotechnic displays ever got up by human skill could rival the exhibitions of nature around our ship." That the cause of this brilliant phenomenon is correctly assigned to marine animals has been proved by the examination of the luminous water, for if it is placed in a tumbler and agitated, they immediately emit light in momentary sparks. Some of these creatures are of considerable dimensions and others barely visible, but a great proportion are entirely microscopic, and require the aid of powerful instruments in order to perceive them and investigate their forms and nature.

The various species of Infusoria which illumine the ocean are extremely small in size; *the largest* do not exceed *one-hundredth of an inch* in extent, while the *least* hardly attain the length of *one twelve-hundredth of an inch.* The phosphoric light emitted by these creatures is regarded by naturalists as the effect of a vital action; it appears as a single spark like that of the fire-fly, and can be repeated in a similar manner at short intervals.

Colored Tracts of the Ocean.—It has been noticed by navigators in all parts of the sea, that extensive tracts of water are not unfrequently discolored at a great distance from land. This change in the hue of the waves is caused by the presence of minute marine animals and Infusoria, which impart their own tint to the waters in which they abound, the far greater part being too small to be observed by the naked eye. Nearly one-fourth of the Greenland sea, comprising an area of more than twenty thousand square miles, is of a deep olive green hue. This coloring matter was discovered by Mr. Scoresby to consist of animalcules, which crowded the water in infinite numbers. On an average, sixty-four animalcules of one kind were found in every cubic inch of water submitted to examination, and, on the supposition that they were equally numerous throughout the body of colored water, Mr. Scoresby computed, that a surface of two square miles, and fifteen hundred feet deep, contained no less than *twenty-three thousand millions of millions* of animalcules belonging to one species. And in order to form a more definite idea of this vast multitude, he remarks that the number of years required for eighty thousand persons to count them, would be equal to the period that has now elapsed since the creation of the world.

This green sea is described as the Polar pasture ground, the animalcules affording an exhaustless supply of sustenance to creatures less minute, and these likewise becoming the food of larger species, which in their turn are devoured by others of greater size. And thus the series continues to increase until the waters are crowded with numerous forms and types of animal life, the prey of the mighty monsters of the deep, which in vast numbers resort to these prolific seas.

On the east coast of Greenland Mr. Scoresby also met with broad patches and bands of water of a yellowish-green color, as if sulphur had been strewn upon the waves; and upon examining it with a microscope it was found swarming with animalcules. Most of them were of a globular form and of a lemon color, and seemed to be possessed of little activity, but the rest were in constant motion. So small were these creatures that the largest did not exceed the *two-thousandth of an inch* in length, and many of them were but half this size. A single drop of the water, and that not the most discolored, was found to contain more than *twenty-six thousand* animalcules. The glass upon which the drop was placed for examination was ruled into small squares of equal size. The drop, when magnified linearly one hundred and sixty-eight times, covered five hundred and twenty-nine of these squares; and in every square, on an average, fifty animalcules were found; which made an aggregate of twenty-six thousand four hundred and fifty. Mr. Scoresby computed that in a tumbler of water *one hundred and fifty millions of these animalcules would find ample room*, and regarding each as one-four-thousandth of an inch long, a row of half a million placed closely side by side, would form a line only ten feet and five inches in extent. Off the coast of Chili, at a distance of fifty miles from shore, Darwin passed in the ship Beagle through wide bands of turbid water; a single tract in one case comprising an area of several square miles. When viewed at a distance the waves appeared red, but under the shade of the veseel, of a deep chocolate color, and the line of division between the red and blue water was clearly defined. Upon close examination in a glass, the water assumed a pale red tint, and when viewed by the microscope was found crowded with animalcules one-thousandth of an inch long, of an oval shape, and encircled at the middle with a ring of cilia. They were beheld darting about in all directions and *exploding*, their bodies bursting to pieces in a few seconds after their rapid motions had ceased. A stratum of red water, *twenty-four miles long and seven broad*, is mentioned by Dr. Pœppig as occurring near Cape Pilares. When beheld from the mast-head it appeared of a dark-red hue, but as the vessel advanced on her course it changed into brilliant purple, while a rosy tint illumined the track of the keel. This water was perfectly transparent, but small red specks could be perceived moving through it in spiral lines.

PROTOPHYTES.

In the various infusions and fluids in which animalcules are found, minute and curious organisms are frequently seen, moving rapidly about and apparently possessed of volition. The earlier microscopists regarded them as true animals, and Ehrenberg arranged them under several families (with numerous subdivisions); such as the Monadinæ or Monads, the Volvocinæ, Astasiaca, Bacillaria, and so forth. These organisms are now, however, considered, by the ablest naturalists, to be plants which rank amongst the simplest and lowest forms of vegetation. For this reason they have received the name of Protophytes, from protos, *first*, and phuton, *a plant*, terms indicating that they are elementary forms of vegetation. These humble types have also been designated as *confervoid algæ.*

The simplest form of plant-life is a cell, and among the protophytes there are many in which every single cell is not only capable of living in a state of isolation from the rest, but normally does so; and thus every cell is to be accounted a distinct individual. There are others again of which masses are made up by the aggregation of contiguous cells, which, though capable of living independently, remain attached to one another by a gelatinous substance which surrounds them. And there are others also, in which a definite adhesion exists between the cells, and in which regular plant-like structures are formed, notwithstanding that every cell is but a repetition of every other, and is capable of living independently if detached. These singular organisms in their development exhibit two marked conditions or states; one is the *still* state, the other the *motile* state.

In the *still* state, the cell for a time remains unchanged, but soon the interior matter separates into two parts, and these again subdivide, and the process is repeated again and again, until sixteen and sometimes thirty-two cells are developed from the original cell. When sixteen cells, and sometimes when even a less number are thus produced by self-division, the new cells assume the *motile* state; for being furnished with one or two threadlike appendages, or cilia, which have the power of producing rhythmical contractions, the cells are thereby impelled through the water in which they live. Erelong they lose their cilia and become invested with a strong envelope, and pass into the still state from which they had previously emerged. This process is continually repeated and causes the plant to multiply with great rapidity.

A class of protophytes, which are termed Diatomaceæ, probably from the readiness with which they are *cut* or *broken through*, are distinguished for the delicate shell of flinty matter which encloses the membranes of the cell.

As the protophytes afford beautiful objects for the microscope, we shall present some of the most interesting forms, and their various modes of aggregation, before proceeding to describe the peculiarities of many of the more curious animalcules. We begin with the *monads.*

MONADS.—These are the smallest of all organisms, which the wonderful power of the microscope has revealed to us. So minute are they, that they must be magnified linearly 300 times in order to be seen at all, and 500 times if we wish to observe them accurately. They appear as transparent globular or oval bodies, moving rapidly about in all directions. Some are of a red hue, others green, many yellow, but the greater part are colorless. All are possessed of one or more of the thread-appendages termed cilia. The monads vary in size, from *one-twenty-four-thousandth of an inch* in length, to *one forty-thousandth* of an inch.

Fig. 8.

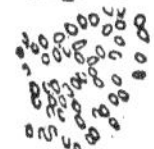

TWILIGHT MONAD.—In figure 8 is shown a group of twilight monads, in which each individual, although exhibited as a mere point, is magnified in length and breadth 800 times, and the space it occupies upon the paper is 640,000 times greater than that which it actually covered in the fluid in which it lived. This organism is globular in form, and presents a glassy appearance. It is found in water containing animal matter; but as the animal substance decomposes the monads unite, forming colorless jelly-like masses, consisting of infinite multitudes of their bodies, which are seen with the naked eye rising and floating upon the surface of the water. This atom is furnished with only a single organ of motion; a delicate cilium issuing from one end, and by the aid of this member it proceeds through the water with considerable rapidity. The twilight monad is only the *twenty-four-thousandth of an inch* long, but it sometimes, though seldom, attains the length of *one twelve-thousandth of an inch*, which it never surpasses. A single shot, one-tenth of an inch in diameter, occupies more space than *seventeen hundred millions* of these structures in their full dimensions, and exceeds in bulk *thirteen thousand millions* of the smallest size. The conception of such minuteness is beyond the grasp of our mind; yet each, an *organized structure*, is not too *small* to claim and receive the regard of Him who called into life and amply endowed it with peculiar organs and properties, adapted to the mode and range of its existence.

Fig. 9.

GRAPE MONAD.—The grape monad is so called from the circumstance that the individuals at times unite and form clusters, like bunches of grapes or berries. A natural group of these protophytes is shown in figure 9, where they are magnified linearly *three hundred and fifty* times; the diameter of the cluster being *one four-hundred-and-thirtieth part of an inch*, and that of each individual *one twenty-three-hundredth of an inch.*

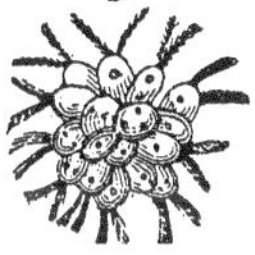

This species possesses an oval form, is furnished with two cilia, and increases by *self-division*, which takes place both across and lengthwise of the body.

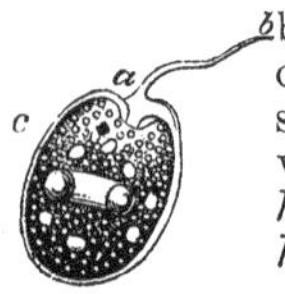

THE GREEN-EYE MONAD.—In figure 10 is delineated a species of the monads in which Ehrenberg supposed that he had discovered a visual organ.

Fig. 10.

It is of an egg-shaped form, and moves in the direction of its length by the aid of a cilium (*a b*), which is nearly as long as the cell. Its color is a rich green, and the eye-speck, which is red, is distinctly seen, as shown in *c*. This organism is found amid water-plants, and varies from *one seven-hundred-and-twentieth* to *one twenty-three-hundredth of an inch in length.* In the figure it is magnified *eight hundred* times in length and breadth.

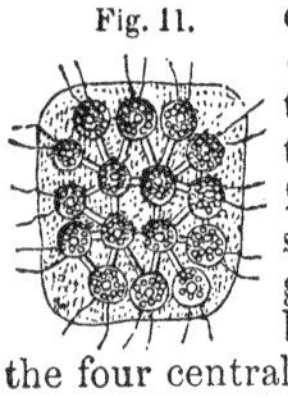

THE BREAST-PLATE MONADS.—Many of the monads are found clustered in one community, and move together as one body. This mode of existence occurs in the breast-plate monads, which have received this name from the form in which they are arranged. A group of these singular plantules is shown in figure 11, and a single one in figure 12. The breast-plate monads are found in clear water, both salt and fresh, and consist of *sixteen globular* bodies of a pure green color, enclosed within a flat, transparent case of a pearly hue. In this they are regularly disposed in a square or oblong form, the four central individuals being usually larger than the rest. Their mode of increase is by self-division, and when this occurs, the group divides across the middle in two directions, separating into four clusters, each containing four monads. No sooner has a group thus separated, than each of the plantules which composed it increases in size, and soon subdivides into four monads, and the original number of sixteen is seen in every one of the four clusters. Erelong, these again separate into four portions, and the species thus multiply interminably. The tablet, though containing sometimes less than sixteen individuals, never exceeds that number. Its form is often irregular; which is caused by the separation of some of the monads from the cluster when they have attained their full growth. Each of the individuals composing the group is connected with the rest by means of six threads or tubes; these, with the two cilia, *a* and *b*, are seen in figure 13, where a monad is exhibited attached to a portion of the transparent case. The length of the tablet is not greater than *one two-hundred-and-eightieth of an inch*, and that of each monad ranges from about *one five-hundredth* to *one-thousandth of an inch.* A single structure, when free from the case, as delineated in figure 12, swims by the aid of its cilia in the direction of the length of the cell, with its ciliated end foremost, as other monads; but the group perform various evolutions, sometimes proceeding horizontally, sometimes upwards, and again rolling on the edge like a wheel. The extraordinary activity of these wonderful little atoms is distinctly beheld when a small portion of coloring matter, as indigo, is introduced into the water in which they are discovered; the whirls and currents will be seen in the fluid, caused by the vibration of the two cilia belonging to each plantule.

Fig. 11.

Fig. 12.

Fig. 13.

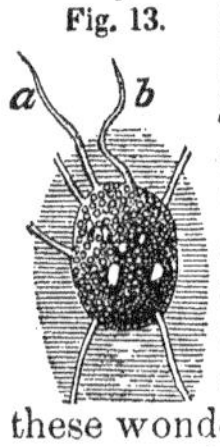

When a group is in progress, thirty-two of these organs are consequently in motion; twenty-four around the edges of the transparent case, and *eight* projecting from the central parts, and, by their combined action, the cluster, enclosed in its delicate envelope, proceeds as one body. Each of the constituent individuals of a cluster is, of itself, a perfect organism, possessing a motion of its own; yet, when united with fifteen others, all act in concert.

THE REVOLVING GLOBE, OR VOLVOX.—About 150 years ago, Leuwenhoeck discovered in water a singular hollow globe, studded with green specks, which advanced through the fluid with a rolling motion. It was at first supposed that the globe was a single animal, but the microscopists of the present day have shown now the true nature of this structure. The little green specks that gem the surface are protophytes; each being a perfect monad, furnished with two cilia, and possessing a bright-red eye-speck. They are all connected together, and every individual is attached to those immediately adjacent by delicate fibres, varying in number from three to six. The thousands thus embedded throughout the entire surface of the transparent spherical shell, form the hollow globe, and bear to it the same relation as the monads of the breastplate cluster to their pellucid case. The whole globe bristles with the cilia of the individual monads; and, by the united action of these slender organs, rolls through the water with the same part always foremost; when the fluid is colored the current and eddies produced by the cilia are clearly detected. This change of place seems necessary for the support of the numerous groups, which range continually in their rolling globe, through new regions of space abounding with matter adapted to their development and increase. In figure 14 the revolving globe is faithfully delineated. The minute dots with which it is covered are the monads that compose it, and the interlacing net-work are the filaments which connect them with each other. The direction of the globe in its progress is indicated by the arrows, and the cilia that propel it are distinctly discerned fringing its surface. Within the globe a number of smaller globes are perceived; and these lead us to consider the extraordinary manner in which these curious groups are multiplied. They increase by a voluntary separation: from time to time new spherical clusters are thrown off from the original globe; not, however, from its outer surface into the surrounding water, but from the *inner surface* into the space enclosed by the transparent shell. Six or eight of these spherical groups are usually found within the parent globe; though, at times, as many as twenty have been seen at once, with their forms well defined, and their color of a bright green. Openings exist, both in the primary sphere and in the interior globes, through which water passes and repasses for the purpose of affording nourishment to these vegetable forms. As the young globes increase in size, the surrounding envelope expands, and as soon as they have attained a certain degree of maturity, it bursts asunder and permits them to escape. Now, uncontrolled in their motions, they range through a wider

Fig. 14.

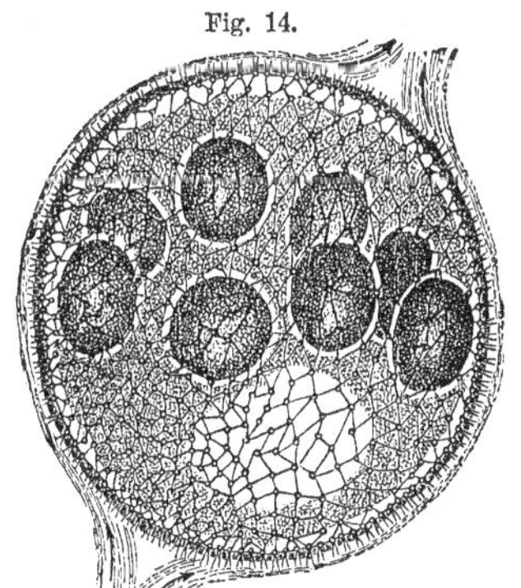

field of existence, and soon a new generation of revolving monads issues from their parting spheres; to become, in their turn, the parents of other globes, and so on in a countless series.

Fig. 15.

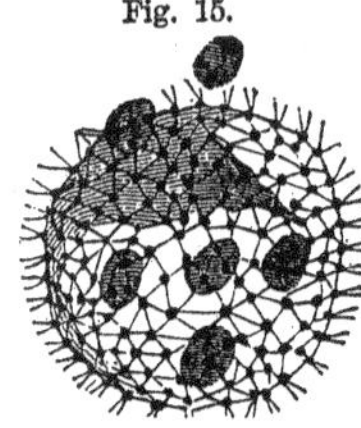

This process of increase is exhibited in figure 15, where the offspring are shown issuing from the parent sphere, and within each of the smaller globes another incipient race of revolving monads is detected. The full sized globes are one-thirtieth of an inch in diameter, and the size of the smallest, when liberated from the parent, is one-three hundred and sixtieth of an inch.

Fig. 16.

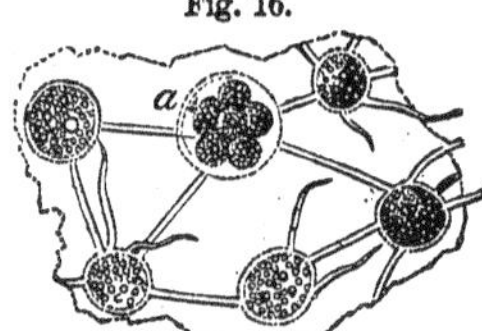

In figure 16 is delineated a portion of a globe with five individuals, and a cluster of six young monads at *a;* they are all attached to the spherical case, and to each other, and the bands which connect them together, as well as their respective organs of motion, are distinctly seen.

Fig. 17.

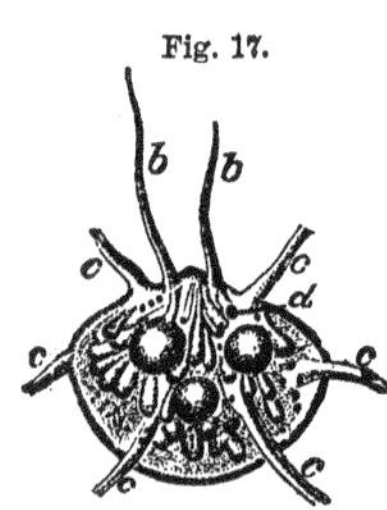

In figure 17 a single monad of a revolving globe, separated from its case, is magnified *two thousand* times; or, in other words, covers upon the paper a space *four million times* greater than its natural extent. In this engraving, the two cilia are seen at *b*, *b*, the six uniting threads at *c*, *c*, *c*, *c*, *c*, *c*, and the eye-speck of the monad, which is of a bright red, is situated at *d*. The natural size of a single organism is the *thirty-five-hundredth part of an inch.* The revolving globe is a common species of protophyte, and is easily found in the clear shallow waters of brooks and ponds.

Ray-Globe Volvox.—Another kind of rolling organisms is delineated in figure 18. It is called the *Ray-Globe Volvox*, and the individuals form, by their union, a group resembling the clustering fruit of the banana. Each organism is enclosed in a cell, and the cells of the cluster are embedded in a jelly-like substance of a spherical form, which rolls through the water like the revolving globe. The ray-globe volvox is of a yellow color, is provided with two organs of motion called cilia; and is likewise furnished with a filament, by which it is connected, either with the centre of the cluster, or with the bottom of its own cell.

Fig. 18.

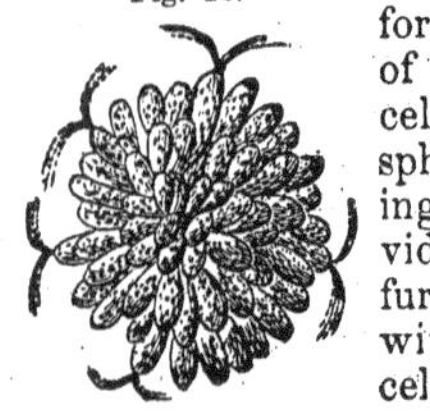

Fig. 19.

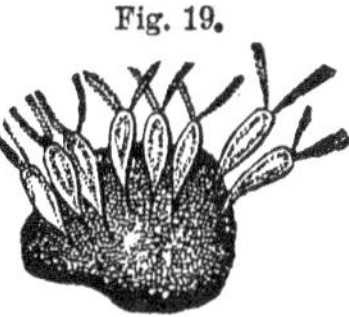

Figure 19 is a magnified portion of a cluster, and displays the manner in which these slender filaments are connected with the common covering. The length of a single organism of this kind, without its filament, is *one seventeen-hundredth of an inch;* and the size of a cluster varies from *one one-hundred-and-ninetieth part of an inch* to *one two-hundred-and-eightieth.*

THE BLOOD-LIKE ASTATIA.—This plantule belongs to a kind which has received the name of Astatia,* from the circumstance that they have no fixed abode like the volvox monads, but are endowed with a perfect freedom of motion. They have the power of changing their form at pleasure, are destitute of an eye-speck, and move from place to place by means of a tail, and a delicate, vibrating cilium. These curious atoms are sometimes produced in such vast numbers as to dye the waters in which they live with a crimson hue.

Fig. 20.

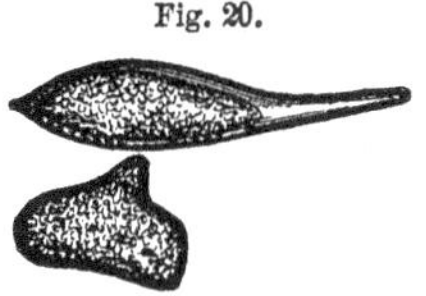

Fig. 21.

The blood-like Astatia is delineated in figure 20. Its body, when extended, is spindle-shaped, as there exhibited; at first it appears of a green hue, but afterwards assumes a blood-red color. Figure 21 shows an individual of the same species with the body contracted. The length of this little organism is *one three-hundred-and-eightieth of an inch.*

THE BLOOD-RED EUGLENA.—The Euglena is a variety of the Astatia, but differs from it in possessing a beautiful red eye-speck. It varies in length from *one two-hundred-and-fortieth* to *one three-hundredth of an inch*, and is of an oblong shape; but is capable of changing its appearance. This curious characteristic is recognized in figures 22, 23, and 24, where the organism is delineated under the various

Fig. 22. Fig. 23. Fig. 24.

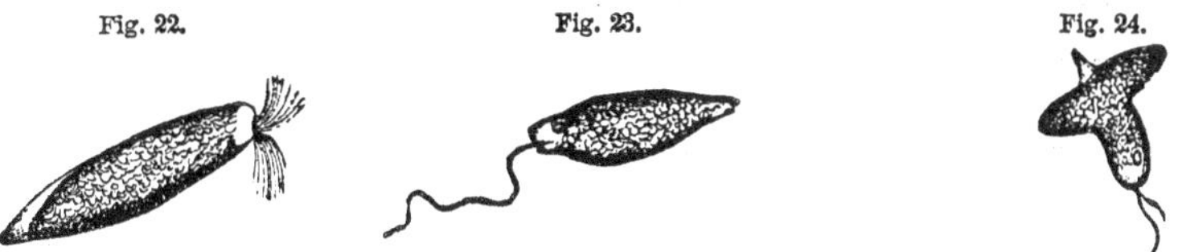

shapes it assumes. During the early stages of its existence its color is green; but upon arriving at maturity it is of a blood-red hue. Individuals are seen, however, partaking of both hues, being variegated with red and green spots. The Euglena moves through the water with a slow motion, by the aid of a thread-like cilium, which is seen in figure 23; and the currents produced by this organ, and which are discernible when the water is colored, are delineated in figure 22. In figure 24, where the cilium appears double, the plantule is on the point of dividing into two, and a single cilium belongs to each of these parts which are soon to become

* Greek, a, privative, *without;* stasis, *a station*, hence Astatia.

independent existences. Not only does this organism swim in a straight line through the water, but it also proceeds on its course by rolling over and over sideways. It is frequently found congregated in vast numbers, clothing with a crimson mantle the surfaces of ponds and stagnant waters.

THE FLOWERING-CUP PROTOPHYTES.—In figures 25 and 26 a singular structure is exhibited, which appears in the shape of a branch; formed of a series of cups united to each other.

The cup is nothing more than a delicate, pellucid shell, enclosing an organism which is attached to the bottom. The living atom, with its encircling case, is distinctly seen in figure 26. It is of a pale yellow tint, and is furnished with a red eye-speck, the position of which is indicated by the oval spot near one end of the organism; and far beyond the margin of the shell protrudes a slender cilium. The flowering-cup protophyte has the power of altering its form, and at one time is seen contracting itself into a round figure at the bottom of its cup, and at another, extending its body as far as the edge of its shell, which is its utmost limit.

Fig. 26.

Fig. 25.

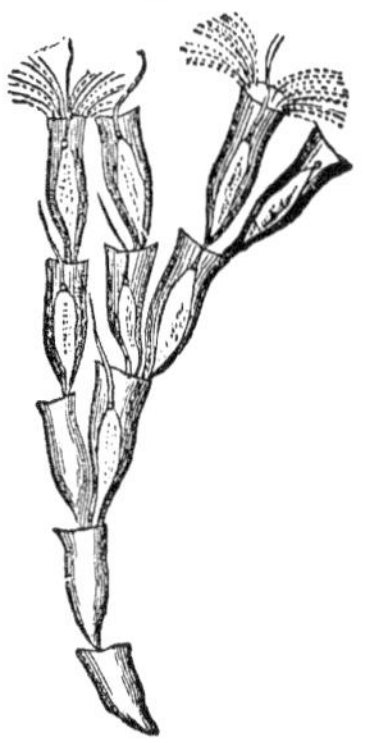

These curious structures multiply by means of little cups, which are seen budding from the parent-cup; and thus it continues to increase, until at length a living branch is developed of considerable size. In figure 25 such a cluster in seen containing eight plantules, and the shells of three which have perished. The motion of the vibrating cilia is indicated by the currents; and through the united and harmonious action of these strange organs, the entire branch of motile organisms proceeds as one body through the water.

This protophyte is found in the water of swamps; its length is about *one-five hundred and seventieth part of an inch*, and that of a cluster *one-one hundred and twentieth of an inch.*

DIATOMACEÆ OR DIATOMS.—These photophytes, which are included by Ehrenberg under the family of the Bacillaria, are invested, as already stated, with a curious flinty envelope which constitutes their chief characteristic. This covering is an object of great interest, not only for the elaborately-worked pattern, which it often exhibits, but also for the perpetuation of the minutest details of that pattern in the specimens obtained from fossilized deposits. The siliceous envelope of every diatom consists of two *valves* or plates, usually of the most perfect symmetry, closely applied to each other like the valves of a mussel. The form of the cavity between the valves differs greatly, each valve being sometimes hemispherical, sometimes like a watch-glass, while in other diatoms they resemble boats, hearts, and so forth. This curious class of vegetable existences are among the most beautiful of microscopic objects.

THE GALLIONELLA.—This division of the Diatomaceæ has been termed Gallionella, from Gaillon, a French naturalist; and has also received the appellation of box-chain organisms, from the form in which they are developed. They are each invested with a flinty case, consisting of two shells; the case being cylindrical in form, and when lying upon its face presenting the appearance of a coin. The cylindrical cases are arranged in chains, in consequence of the imperfect self-division of the cell, whereby the young forms, as they are successively produced, remain attached to the parent stock. The Gallionella are found both in a living and in a fossil state, and in the latter afford very rich objects for the microscope. They are exceedingly abundant, existing in every pool, river, and lake; and such are their astonishing powers of increase, that *one hundred and forty millions of millions* will spring from a single specimen, by self-division, in twenty-four hours.

Fig. 27

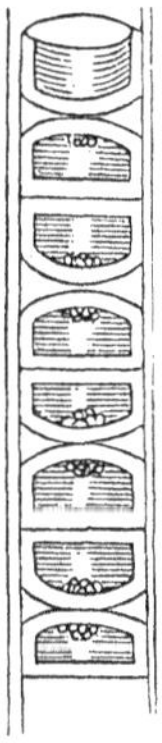

THE STRIPED GALLIONELLA.—This species of the Gallionella is delineated in figure 27, which represents a specimen found by Dr. Mantell, in a pond near London. Several distinct diatoms, invested in their flinty cases, are here beheld forming a chain, which is highly magnified. The fine lines running across the various links of the chain are in the direction of the length of the organisms, and the position within them of masses of green and yellow atoms is indicated by the small circles distributed throughout the entire chain. The striped Gallionella is found both in fresh and salt water, and various in size from *one-fourteen hundredth* to *one-four hundred and thirtieth of an inch.* The latter length is the natural size of the engraved specimen. Single chains are sometimes found three inches in length, consisting of from 1,200 to 4,000 organisms.

THE RUST-LIKE GALLIONELLA.—In figures 28 and 29 are shown several delicate and branching chains, composed of these little plantules, which, from the resemblance of their color to that of iron-rust, have received the name of *Rust-like Gallionella.* Each of these individuals is invested with a flinty shell, of an oval shape, rounded at both ends and smooth on the surface. They are found in most of the waters impregnated with iron, and also in peat-water, which contains a little of this mineral. Every thing beneath the surface is covered by these existences in countless numbers, forming, by their union, a light mass composed of such delicate flakes, that it is dispersed by the slightest motion. In the spring, this flocculent substance consists of short chains of pale-yellow globules, strung together like rows of beads, which can be readily separated from one another. Their union, however, is detected with difficulty at this time; but as the season advances the diatoms become more developed, and their structure is better discerned: when summer arrives, the threads and chains of which the whole mass is woven, yield to the power of the microscope, and the texture is more clearly revealed to the eye. At this period, the color of this organism is of a deep rusty red; but in the spring, its tint is that of a pale-yellow

Fig. 28. Fig. 29.

ochre. This species of Gallionella is found both in a recent and fossil state, and measures only *one-twelve thousandth of an inch in diameter.*

THE NAVICULA.—The Navicula, both recent and fossil, forms a very interesting class of the diatoms; it is shaped like a boat or ship, and from this resemblance has received the name of Navicula, which in the Latin language signifies a *little ship.* They are never united, like the Gallionella, in chains; but exist singly and in pairs, enclosed in a durable, thin, siliceous cell, generally four-sided, and which, when slightly pressed, divides either into two or four parts, disclosing the appearance of ribs running across it. A jelly-like substance, which constitutes the body of the plantule, occupies the interior of the shell, and portions of matter, of a green, yellow, and brown color are here perceived, which were once regarded by naturalists as the eggs of the Navicula. Many of the Ship-Navicula propagate by self-division in the two directions of their length and breadth; the separation commencing in the soft body beneath the shell, which afterwards divides into parts corresponding to those of the body. Twenty-four different kinds of fossil Navicula have been discovered, fourteen of which have been identified with species now living.

GREEN NAVICULA.—Figure 30 is a drawing of this diatom, representing a specimen taken by Dr. Mantell from a pool in Clapham Common, in the vicinity of London. The small dots which Ehrenberg supposed to be stomach-cells, and the rib-like divisions of the shell, are distinctly seen throughout its whole extent. So numerous are they, that fifteen are contained within every twelve-hundredth of an inch in length. A side view of the same individual is shown in figure 31, exhibiting the currents produced by its motion through the water. This species of Navicula varies in size from *one-seventieth* to *one-one hundred-and-fifteenth part of an inch in length.*

Fig. 30.

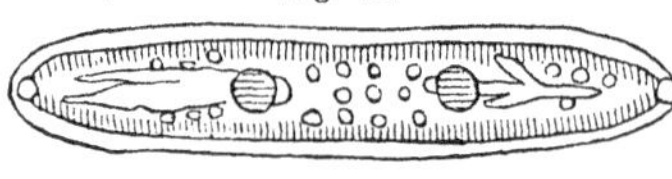

Fig. 31.

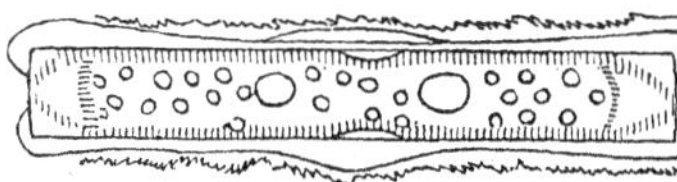

GOLDEN NAVICULA.—Figures 32 and 33 are representations of a beautiful species of the golden Navicula, so called because the clusters of globules within the shell are of a bright-yellow color. They are seen in the engraving occupying the central portions of the shell, and filling up its numerous flutings. The shell is of an oblong oval shape, and possesses the utmost regularity in its structure. In figure 31, the Navicula is seen from above; in figure 33, a side view of the same diatom is presented: by comparing the two drawings, it is seen that the shell tapers more in the latter case than in the former. The above figures are faithful delineations of a living golden Navicula obtained by Dr. Mantell: the

Fig. 32.

Fig. 33.

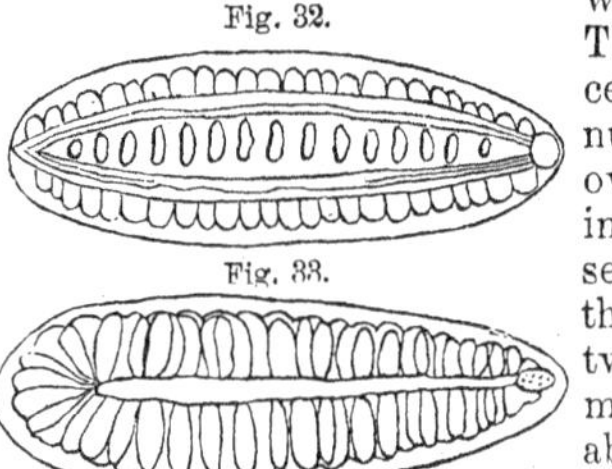

length of this specimen was found to be the *one hundred and forty-fourth part of an inch.* This species varies in size, however, from *one-one hundredth* to *one-two hundred and tenth of an inch.*

The Swollen Eunotia.—In figure 34 is shown the shell of a species of diatomaceæ which differs a very little in its characteristics from those just described.

Fig. 34.

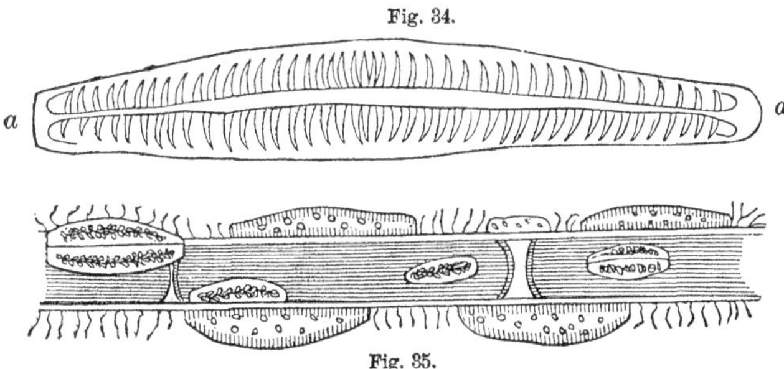

Fig. 35.

It is called the Swollen Eunotia. The shells vary in length from *one-eleven hundred and fiftieth of an inch to one-two hundred and fortieth,* and are of the shape represented in the figure, which exhibits a side view. A furrow (*a, a*) runs the whole length of the shell, along the middle of each side, and from this furrow numerous curved ribs branch out towards either edge. The furrows are plainly

Fig. 36.

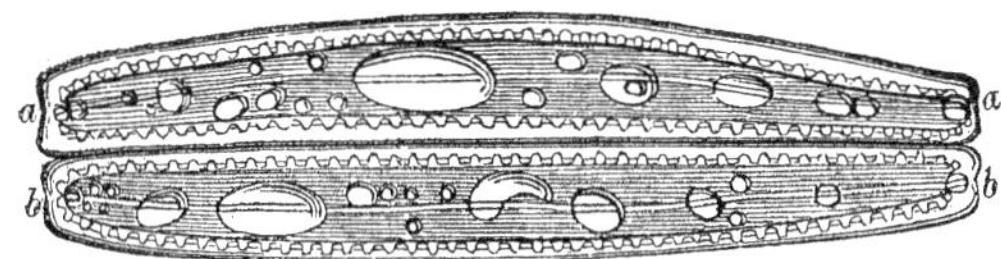

discerned in the shell; but are detected with difficulty in the living organism, on account of the color of the body. So closely are the ribs placed together that no less than eight are contained within the space of *one-twelve hundredth of an inch.* Figure 35 is the representation of several living individuals of this species, found upon a branch of conferva, which is the bright green vegetable matter that floats upon stagnant waters during the spring and summer. The Eunotia multiplies by self-division, and in figure 36 an individual is exhibited undergoing this process. The separation is seen to take place in the direction of the length, and in each half we can discern another line of division (*a, a; b, b*) just commencing. Through this line, when the divided portions have arrived at maturity, and each has become a perfectly developed plantule, another separation occurs, and thus proceeds interminably.

THE PYXIDICULA.—These minute organisms have received their scientific name from their form—*pyxidicula* signifying, in the Latin language, a *little box.* They are enclosed in a transparent, spherical, flinty case, which is marked by a circular furrow through which it readily divides, separating into two hemispheres. A group of a living species of the Pyxidicula is delineated in figures 37, 38, 39: *a* is a view of the shell at right angles to that presented at *b*, and exhibits the furrow through which it separates; and *c*, is one of the two hemispheres into which the shell divides. This organism is of a yellowish-green color, and varies in length from *one-fourteen hundred and fortieth of an inch* to *one-five hundred and seventieth.* It is quite common, and is found both in a living and fossil condition.

Fig. 37. *a* Fig. 38. *b*

c Fig. 39.

Fig. 40.

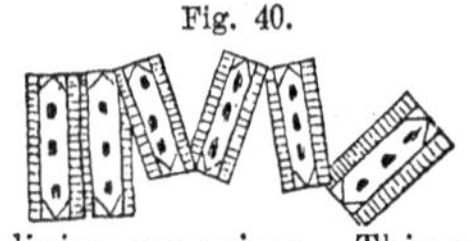

THE ZIGZAG DIATOMS.—Figure 40 is a drawing of a common Zigzag diatom, found by Dr. Mantell in the neighborhood of London. These protophytes have received the above appellation in consequence of their being developed in zigzag chains, each link consisting of a living organism. This mode of union arises from the circumstance that, although the shells of all the diatoms are perfectly separated, their bodies are not, and thus remaining attached, they present to view an irregular series, such as is displayed in the figure. The flinty shell is three or four times as long as it is broad, is prismatic in shape, and contains thirteen cross lines in every twelve-hundredth of an inch. A narrow opening runs from one end of the shell to the other, through which certain processes are protruded, by the aid of which locomotion is effected. The natural size of each shell of the chain in the engraving is *one-four hundred and thirty-second part of an inch.*

Fig. 41.

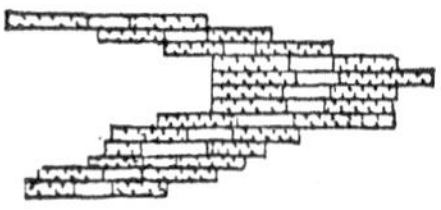

In figure 41 is shown a cluster of another species of these diatoms, which, when imperfectly divided, are attached side by side, and slide one upon the other—the entire group shortening and lengthening itself at, as it were, at will. Their color is of an orange yellow, and their length varies from *one-two hundred and fortieth of an inch* to *one-eleven hundredth.*

THE PALM, FAN-SHAPED DIATOM.—A species of protophytes belonging to the type of the Diatomaceæ, and bearing the above name, is shown in figure 42. These organisms are encased in a shell which is broad and wedge-shaped, and form a fan-like cluster, rising from a single trunk or stalk. The stalk is produced by an excretion of the plantule, and is not possessed of any vital power; for if the branching, living groups are broken off, no fresh buds, teeming with vegetable life, are put forth from the mutilated trunk, but it soon crumbles away and utterly perishes. This organism increases by a longitudinal self-division, but the separation does not extend to the stalk; for this remains entire while the diatom continues to develop in fan-shaped groups, the trunk branching into thick gelatinous boughs, to which the separate vegetable atoms, gleaming with a golden hue, are attached like fruit by a slender stem. This organism is found covering the surface of marine plants. The natural size of the cluster varies from *one-twelfth* to *one-sixth* of an inch—that of a single specimen is *one-one hundred and twentieth of an inch.* Figure 43 is a back and side view of a single individual, highly magnified.

XANTHIDIA.—The Xanthidia belong to the Desmidiacæ, one of the simplest species of the Protophytes; it is enclosed in a transparent, single-valved shell, of a globular shape, which resists the action of fire, and is studded with spines or thorns: a green mass is seen in the interior of the body which was once supposed to consist of eggs. The Xanthidia exist both in a living and fossil state, and are found abundantly in flint, as will be shown hereafter. They exist singly, in pairs, and in groups of four, and increase by self-division. Two figures of living Xanthidia are displayed in figures 44 and 45. Figure 44 is a drawing of a forked Xanthidia, found by Dr. Bailey in a pond near West Point: its shell is green and of oval form, and its natural length is *one-two hundred and eighty-eighth of an inch.*

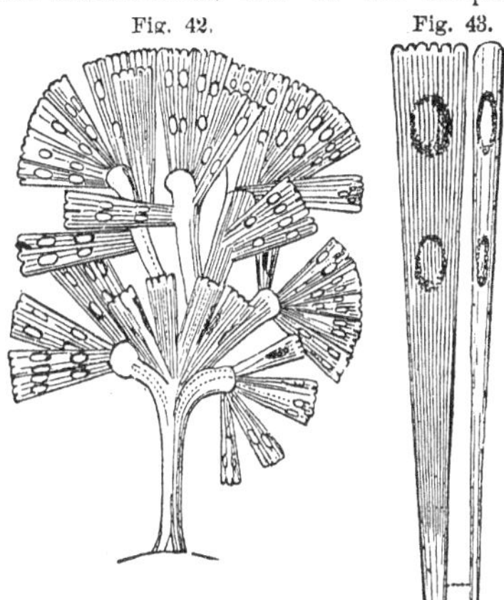
Fig. 42. Fig. 43.

Figure 45 is a different species, and represents a spinous Xanthidium, obtained by Dr. Mantell from a pond in Clapham: it is of the same size as the preceding specimen, and is likewise of a beautiful deep-green hue.

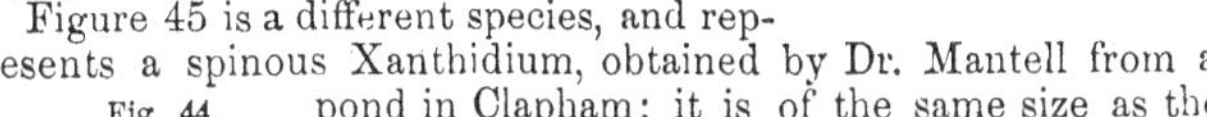

Fig. 44.

Fig. 45.

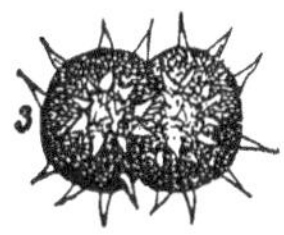

POLYGASTRIC ANIMALCULES.

THE PROTEUS.—In figures 46, 47, and 48, a most remarkable animalcule is exhibited, which varies in size from *one-one hundred and fortieth* to *one-seventieth of an inch* in length. It appears under the microscope as a pale-yellow mass of jelly-like matter, and is endued with the power of changing its shape to a very extraordinary degree, as is obvious from the inspection of the figures. From this circumstance it is termed the Proteus, the name of the wondrous sea-god, who could assume, at will, every form, turning himself into animals, trees, fire, and water, according to the fables of the classic poets.

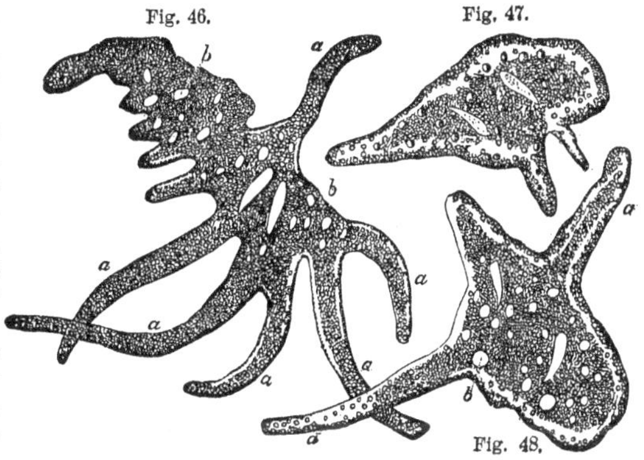

Fig. 46. Fig. 47. Fig. 48.

The Proteus can hardly be said to possess any original shape; for it is capable of relaxing itself in one place, and contracting in others; and of pushing out from every part of its body long arms and feelers (*a a a*, &c.), which are its organs of motion, to the number of ten or twelve at one time. These members the animalcule can again withdraw into its body, and protrude others from a different place, if it pleases so to do. This animalcule, though classed with the Polygastrica, can hardly be said to possess stomachs, for it is simply a mass of matter composed of cells. Within this, numerous contractile vesicles are seen which were, at first, supposed to be stomachs.

In the figures, their situations are indicated by the larger cavities (*b b b*, &c.), and are represented as they exist, dispersed throughout the body of the creature.

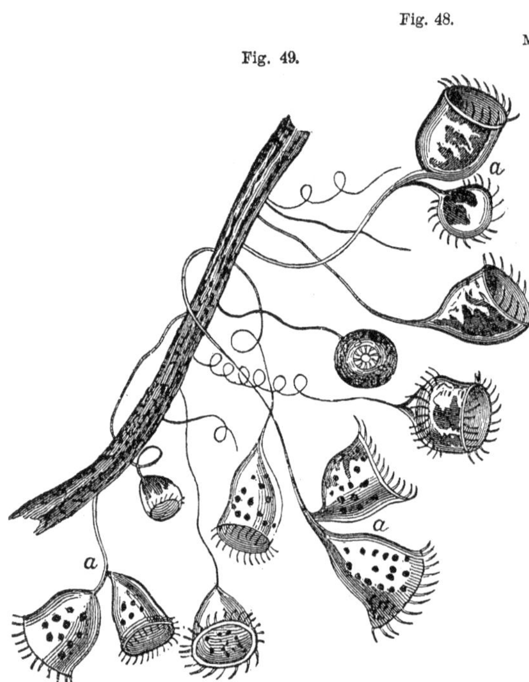

Fig. 48.

Fig. 49.

THE BELL-SHAPED ANIMALCULES.—This family of Infusoria, which is remarkable for the graceful elegance of its forms, is devoid of a shell; and each individual, when unconnected with others, roams about in solitary independence. When, however, they are attached to a stem, they live together in great numbers, assuming the shape of trees or shrubs, with an animalcule appended like a flower to the extremity of every tiny spray. Imperfect self-division gives rise to these beautiful tree-like clusters; but in addition to this mode of increase, they likewise multiply by the growth of buds, either from the sides of the animalcules, or from the stalks to which they are united. The Bell-shaped animalcules are usually found clustered together, in countless numbers, upon the submerged surfaces of twigs and roots. They adhere also to the small leaves of the duck-weed, and attach themselves to the shells of minute aquatic animals; but when fully developed, they are generally connected with some fixed object. A group of several animalcules belonging to this family, and of the species termed the Nebulous bell-shaped animalcule, is exhibited in figure 49. The body of the creature, as its name implies, has the shape of a bell, the margin of which is fringed with a circle of cilia. The space surrounded by the cilia is the mouth of the animalcule, and the position of its stomachs is marked by the round spaces within the bell. The slender stem by which each individual of a group is attached to a common base is furnished with a long and delicate muscle, traversing its entire length, by the aid of which it is alternately extended and shortened. The animalcule, at the approach of danger, coils up its tiny cable with the quickness of thought, sinking down towards the spot where it is anchored; and then again when the peril is past, floats upwards in search of food, and swings once more to the utmost extent of its line. This contractile action, as Mantell remarks, is continually going on—now, in one or two individuals only, then in several, and often the whole group suddenly shrinks down into a confused mass, and the next instant expands, and every little bell becomes fully developed, with its cilia in rapid oscillation.

The group in the figure represent a number of animalcules, of this species, in different attitudes and conditions. Some have their stalks stretched to the utmost length, with their crowns of cilia in full action, while others have their stems more or less coiled. Many are single, and others (*a a a*,) have divided into two, and the stems of several are seen, from which the animalcules have broken and swam away. The length of the body of this species varies from *one-two hundred and eightieth of an inch* to *one-five hundred and seventieth.*

The self-division of these Infusoria proceeds as follows: The bell first begins to expand in breadth, and then a separation commences extending in the direction of its length, double rows of cilia becoming meanwhile formed. At length, the two parts being perfectly developed, the bell divides into two animalcules, and fringes of cilia next appear encircling the base of each. Soon the young animalcules twist off from the stem which speedily decays, and after swimming about for some time, at last put forth a new stem from the end of each bell; then fixing themselves to some object, they multiply by self-division, and become in their turn the progenitors of a numerous race.

Fig. 50.

In figure 50, two bell-shaped animalcules are seen, which have just been produced from one by separation, each being still attached to the parent-stem by its own stalk.

The advantage resulting from the use of a colored fluid, in enabling the observer to detect the stomach-cells of minute Infusoria, has already been mentioned more than once; but the following detail of its employment by Dr. Mantell, when experimenting upon Bell-shaped Infusoria, is too instructive to be omitted. "I place," says this interesting writer, "a drop of a solution of carmine in the water between the plates of glass containing these animalcules; the fluid in which they are floating now appears turbid, and full of gray particles, which are thrown into rapid motion by the vibrations of the cilia, and currents are seen passing to and fro from the mouths of the animalcules. In a few minutes the water gradually becomes clear, and several round spots of carmine are apparent in the body of each animalcule. We have, in fact, caused their little stomachs to be filled with coloring matter, and can now distinguish their number and arrangement. If the body of the animalcule were a mere cavity, it is obvious that the carmine must have collected into a single ball or mass; but this is not the case ; on the contrary, the color appears in distinct round spots, from having accumulated in globular cells, and, by careful investigation, the tube connecting these cells may be detected."

THE TREE-ANIMALCULES.—The Bell-shaped Infusoria, which have just been described, are each attached to a common base, by a separate stalk ; but in the kind we are now considering the entire cluster springs from a single trunk, which, dividing and subdividing into numerous branches, puts forth its trumpet-shaped living blossoms, at the extremity of every bough. This mode of union is the result of imperfect self-division—a single animalcule first separating into two, that are united by a forked stem, and these again into four, which still remain connected ; thus the division proceeds, until the limit of development is attained, and a graceful, branching cluster rises from a common stalk. The spontaneous separation of the individual creatures is effected in the manner already detailed, and, as in the case of their kindred species, the Tree-animalcule, upon arriving at maturity, breaks away from the parent stock, and the single living cup, floating for awhile, at length becomes stationary, and a new generation of arborescent forms are produced, by their repeated self-division. These Infusoria have also been observed to multiply by the growth of buds.

The Tree-animalcule is shown in figures 51 and 52. In the latter engraving the beautiful group is seen fully developed, with the trumpet-shaped animals clustering on every branch ; their circles of cilia expanded to the utmost, and the position of their stomach-cells clearly indicated by the round spaces within their bodies. A muscle is distinctly seen rising from the root, traversing the trunk, and extending its minute ramifications to every member of the group. Not only is each individual endowed with the power of coiling and uncoiling its own stalk, by the aid of this muscle, independently of the action of its fellows, as shown in the cut, but the whole group can suddenly contract its dimensions, the main trunk folding closely together in spiral wreaths, and the spreading tree shrinking into a globular mass, while each animalcule assumes a spherical shape, and its crown of cilia occupies a narrower circle. This attitude of the group is exhibited in figure 51. The natural size of a single animalcule ranges from *one-four hundred and thirtieth of an inch* to *one-five hundred and seventieth.*

Fig. 51.

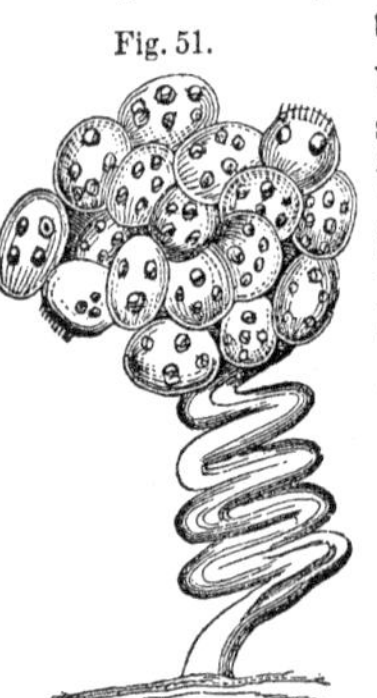

The living forms that constitute the Infusorial world, bear, for the most part, little or no resemblance to those that are visible to the unaided eye ; but in the beautiful groups we have just considered it is otherwise ; for, in their curious figures and organization, that type of animal existence is recognised, which adorns the ocean with living flowers, and peoples its azure depths with a thousand arborescent forms ; where a sensitive life is enshrined in each bud, and which, developed in its countless generations, spreads through the waters, a mazy grove.

TRUMPET ANIMALCULES, (*Stentors.*)—This division of the family of the Bell-shaped Infusoria receives its name from its peculiar form, which resembles that of a trumpet. Unlike most of the race to which it belongs, the Stentor is destitute of a stalk, and attaches itself by the lower extremity, in the manner of a leech, to the different substances it meets with in the water. Its whole body is covered with cilia, and a spiral wreath of these organs is seen surrounding the expanded mouth. By the aid of these members the Stentor moves swiftly through the water, and at the same time sweeps within its grasp the various living atoms upon which it preys. This creature is very voracious and devours great quantities of monads, wheel-animals, and other Infusoria. These are frequently found within its stomachs, which are arranged like the beads of a necklace. This chain of stomachs proceeding from the mouth, traverses the body of the Stentor in the direction of its length, and returning, unites with it in a spiral-shaped cavity. They increase by self-division, either lengthwise or obliquely, and also from eggs which vary in color in different kinds. There are several species of the Trumpet-animalcules, which are dissimilar both in size and color. Some of them are sufficiently large to be detected by the naked eye, but the microscope is needed for their full examination. Their colors are blue, vermilion, green, yellow, and brown. In figure 53, a group of a species, termed the *Many-shaped* Stentor is delineated, adhering to a stick. Four individuals are here seen, gracefully entwined together in various attitudes and in different states of expansion. In the upper figures, the distended mouth encircled with its spiral wreath of cilia is fully displayed, and numerous cilia are likewise seen covering the surface of the body, but appearing most thickly near its lower extremity, by which the animalcule has anchored itself to the sunken spray. In the remainder of the group the mouths of the animalcules are turned from view, but the position of these orifices is marked by the surrounding crowns of cilia. These animalcules are of a beautiful green hue, and vary in length from *one-twenty-fourth* to *one-one hun-*

Fig. 52.

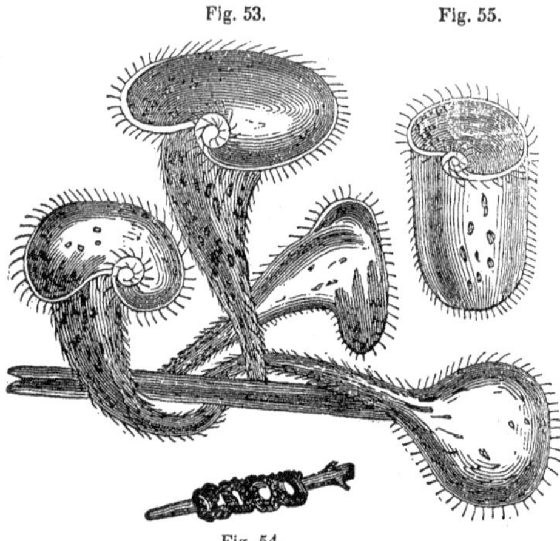

Fig. 53. Fig. 55. Fig. 54.

dred and twentieth of an inch, and are found abundantly in stagnant water, upon decayed sticks, stones, and leaves; on the surface of which they cluster in countless myriads. The appearance they then present is similar to that shown in figure 54, which represents a twig encased in a mass of green jelly, consisting of thousands of animalcules of this kind. The twig was taken from a lake by Dr. Mantell, and was part of a branch three feet long, that had fallen into the water and was entirely covered with the congregated multitudes of these Infusoria. The

Fig. 56.

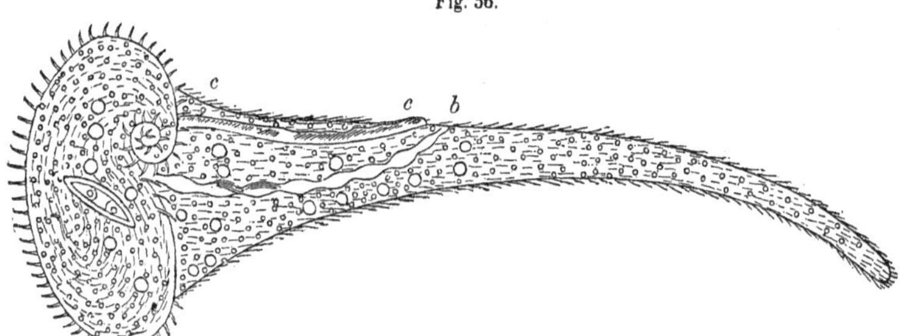

Fig. 57.

group just described displays the animalcule when extended to its full length; but in swimming it contracts into the shape of a cylindrical cup with a spiral margin, as exhibited in figure 55. The *Blue stentor* is shown in figures 56 and 57. In the first it is seen elongated, as it appears when attached to some object, and in the latter under the shape it assumes when swimming. Its voracity is plainly evinced by the number of animalcules within the enclosure of its funnel-shaped mouth, below which the rows of stomach-cells are apparent, extending to *b*. This species has a crest (*c*, *c*,) extending along its body. The length of the creature is *one-four hundred and eightieth of an inch.*

Purse Animalcule.—A species of this animalcule is represented in figure 58. Its body, which is white and round, is covered with cilia, arranged in circular rows, and the double fringes that surround the large opening which constitutes its mouth are longer than the rest. Its stomachs, which resemble in shape small purses, are not connected together in a chain; but are attached to the interior of the animalcule, by slender stems. These Infusoria are found with the dust-monad and tablet-animalcules, which they devour in great numbers, and in the specimen delineated, several of these creatures are seen within its body. The natural length of this species of purse-animalcule is *one-one hundred and eighth of an inch.*

Fig. 58.

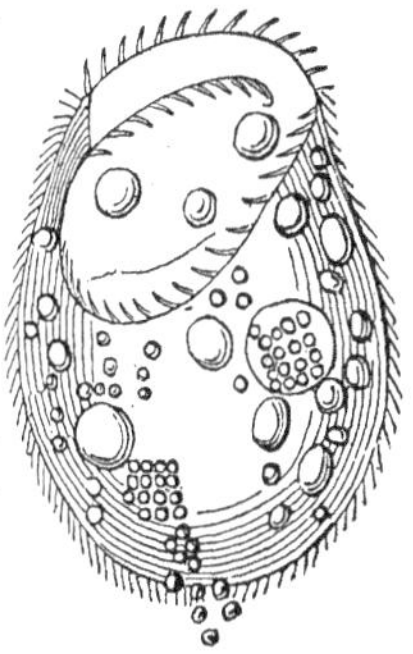

WHEEL-ANIMALCULES, OR ROTATORIA.

These interesting animalcules, of which there is a great variety, constitute one of the great classes into which the Infusorial world is divided. They live, for the most part, in water; but it does not appear to be necessary to their existence that they should be enveloped by this fluid. They are frequently found to reside in moist earth, and some species are known to dwell in the cells of mosses and sea-weed. We have already observed that these Infusoria have received their name from the apparent revolution of their crowns of cilia, and that in addition to this marked peculiarity, they are distinguished from the Polygastric animalcules by possessing a single stomach, and in being furnished, for the most part, with jaws and teeth. The Rotatoria, though endowed with the power of changing the shape of their bodies, by contraction and expansion, cannot do so by the growth of buds, or self-division, like many kinds of the first class. At the lower extremity of the body is a short stem, or tail, by which the animalcule fastens itself to some fixed object, at its pleasure, and thus prevents the upper part of the body from partaking of the motion of the cilia. This class of Infusoria are *viviparous,** and also multiply from eggs, which, in some species, are equal in length to one-third of the extent of the body. For an unknown period of time, the eggs retain the living principle within them, exposed to heat and cold, to moisture and to drought: they are borne upon the wings of the wind, over sea and land, bursting into life, and filling the water with their swarming multitudes, whenever a concurrence of circumstances favorable to their development calls them into existence. Nor is this tenacity of life confined to the egg; the animalcule itself, as we have previously shown, slumbers for months, in apparent death, and is repeatedly revived.

* Producing young in a living state.

These singular creatures are found in the red sediment left in gutters and troughs, after the rain-water contained in them has evaporated; also in vegetable infusions, especially that of hay; and they likewise swarm in great abundance in ponds covered with water-plants. In sea-water they are also detected, and it is said that they have even been observed moving freely in the cells of terrestrial and marine plants.

The wheel-animalcule enjoys the sunshine, and can seldom be taken in a cloudy day; for then it seeks the bottom of the water, lurking around the roots of the weeds that grow therein. When the small pools in which they reside have been reduced by evaporation, they become so numerous as to tint the water with a bright red. They have then attained their full size, and the richness of their hue is at its height; but if they are now confined in a vessel for a few days, their color fades entirely away. Pritchard remarks that he had found wheel animals, taken under the circumstances just detailed, to measure *one-thirtieth of an inch* in length; while those raised in artificial infusions seldom exceeded half that size. They were so numerous that thirty were contained in a single drop. He also observes, that they are easily preserved for a long time, by occasionally placing a little hay in the water of the glass vessel in which they live. He was enabled in this manner to keep them for the space of five years; and the drawings we shall now describe are representations of specimens preserved. The wheel-animalcules are mostly seen under the forms delineated in figures 59 and 60—the first of which represents a full-grown animalcule, and the second the same when young. Cup-shaped, wheel-like organs surmount the head of the creature, and are each furnished with circles of cilia, apparently in constant rotation; causing currents, as shown by the arrows, to set towards the opening between the wheel, bearing along the particles of matter upon which they feed. From this opening the food is carried through the neck to the mouth, situated at the bottom of the neck. These curious organs are more clearly displayed in figure 7, where not only the crowns of cilia, but the jaws, teeth, and eyes, are delineated as they appear when very highly magnified. Here the rotatory organs are seen consisting each of twelve or fourteen groups of cilia, which, swinging round upon their bases, describe small conical surfaces, terminated by the dotted circles. The animalcule is endowed with the power of changing the direction in which the wheels appear to revolve, and also of instantly drawing in the whole of its wheelwork. The head then assumes the form presented in figure 61, where it is terminated by a cluster of hairs that do not revolve. This tuft is regarded as a set of filaments distinct from those that encircle the wheel-organs; and which are supposed to perform the office of feelers, as they are usually protruded when the animal is moving from place to place. Near the head, and at the upper part of the tube through which the creature receives its food, are four muscular masses of a hemispherical form, placed opposite to each other, as shown in figure 7. Two of the masses are furnished with jaws and teeth; the jaws are semi-circular in form, and are each armed with two teeth, that are inserted in the arched portion of the jaws. These organs are frequently seen in action, when the creature is feeding, and are distinguished without difficulty. Between the rotatory organs are situated the eyes of the animalcule, which are two in number, and of a red hue. For certain reasons, Dr. Ehrenberg has been led to suppose that these eyes are not *simple*, but *complex*, like those of insects; each organ of vision consisting of a number of lenses, which form as many separate images of a single object before them There is also a tube projecting from the neck, the position of which is indicated at *b*, in figure 61; through this, water flows into the body of the creature, for the purpose of affording constant supplies of air. The wheel-animalcule moves through the water by two different methods. The first is by *swimming*, which is accomplished by the rotatory action of its crowns of cilia; and in the second method the tail is employed. This member is provided with two pairs of projections, *g g*, figure 59, which may be termed feet, and is likewise divided at the end. By alternately attaching its head and tail to the surface of the object upon which it moves, the animalcule advances in its course, bending itself upward in the manner of a caterpillar, in order to effect this object. Wheel-animals progressing in this way are delineated in figures 62 and 63. In the various figures presented, we perceive joints and rings, like those which surround the common worm. The joints are not limited in number, nor confined to any particular situation on the body of the animalcule, and where any joint occurs the smaller parts slide in and out of the larger, like the tubes of a telescope. From this peculiarity these Infusoria can assume the form of a sphere, the head and tail being drawn within the body. This movement is nearly effected in figure 64, the toes being still attached to a stem. In figure 65 the animalcule is entirely contracted, and forms a spherical ball. A number of the eggs of the wheel-animalcule are shown below figure 59. In form they are oval, with a richly granulated surface. They vary in color, being sometimes of a delicate pink, and at others of a deep golden yellow.

The Crown Wheel-Animalcule, or Stephanoceros.—This elegant little creature has received the name of *stephanoceros*, from the Greek word stephanos, a *crown*. It is found in ponds, amid the leaves of aquatic plants, and usually measures one thirty-sixth of an inch in length. A specimen of these Infusoria, which was carefully studied and delineated by Dr. Mantell, is represented in figure 66, and displays at a glance its singular peculiarities of structure. The Stephanoceros is here beheld fully extended, enclosed in a transparent, cylindrical case, *a a*, which is flexible in its nature, and is attached to the body of the animalcule, near the head, as seen at *c*. The head is adorned with a crown, consisting of five long branching arms, *d d d d d*, each of which is fringed with fifteen small circles of cilia, that are perpetually vibrating. Upon the currents produced by the motion of the cilia, the prey of the Stephanoceros floats, till it comes within its grasp, when it is seized by the long arms of the creature, and firmly held until it is devoured. In the figure several animalcules

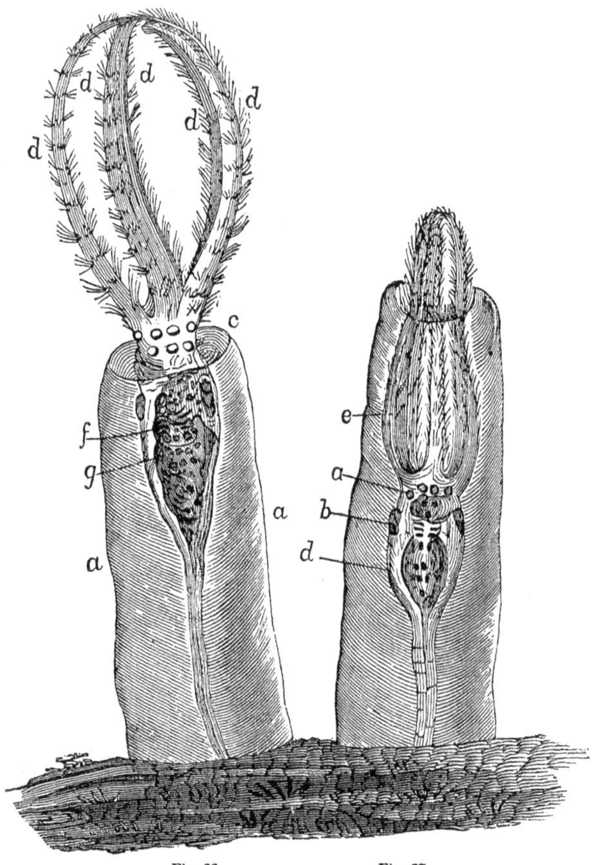

Fig. 66. Fig. 67.

are seen which have thus been captured and swallowed: it is likewise by means of these fringed arms that locomotion is effected. In figure 67, the Stephanoceros is seen drawn within its case. This change of attitude it effects instantaneously, upon the slightest alarm: the arms are then closed together, and the case contracts in wrinkles; the upper edge being drawn inward, as the part of the animalcule to which it is united sinks down toward the bottom of its crystal cell. The tube is thus *doubled inward* upon itself, like the finger of a glove turned partially outside in.

The Stephanoceros is said to possess a single, small, red eye, which is not seen in these figures. The minute circles at *c* represent several fleshy masses, which are arranged two and two at the base of each arm, and are supposed to be centres of the nervous matter provided for each member. The position of the mouth is indicated by the letter *f*, and below it, at *g*, is the stomach, which is comparatively large, and is here exhibited filled with many of the smaller Infusoria. The jaws of the Stephanoceros are furnished with teeth, with which it is seen to tear and masticate its food. Two distinct sets have been discovered on each jaw; of which, in their natural position, figure 68 is a highly magnified representation. The lower set is seen at *a*, and the upper at *b*; the latter appear to consist of two on each side; but they are not all seen in the figure, for the actual number is eight, four upon either jaw. So fierce and voracious is the crowned animalcule, that it attacks and seizes the Stentor with its long and flexible arms.

Fig. 68.

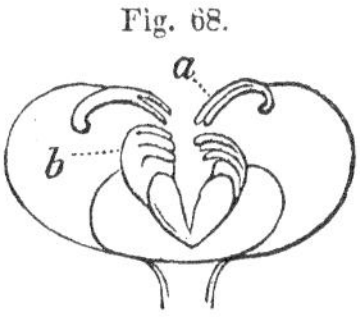

The Stephanoceros increases by eggs, which are hatched before they pass from the animalcule into the cavity of its transparent case. The progressive development of the young, during the first stages of their existence, has been studied by Dr. Mantell with the most patient assiduity. This gentleman observed that the young Stephanoceros, three hours after it had escaped from the egg, swam freely in the surrounding water; in thirty hours a group of five buds were beheld, which were regarded as the bases of the five branching arms constituting the crown; in eighty hours they were fringed with cilia, and the position of the stomach was detected by the color of the food which the young animalcule had swallowed.

The specimens of the crowned animalcule, which are represented in the above figures, belong to a species obtained by Dr. Mantell from a lake in the vicinity of London. A supply of them was kept without any difficulty in glass jars of water, containing aquatic plants, during the residence of this gentleman at Clapham. But upon removing to another place, these interesting creatures died, although they were furnished with their native water, and every precaution was taken to ensure their lives. This mortality is attributed by Dr. Mantell to a difference in the local influence of the atmosphere.

THE BEADED MELICERTA, OR FOUR-LEAVED ANIMALCULE.—This minute creature, which is delineated in figures 69, 71, and 72, belongs to the same family of Infusoria as the Stephanoceros. It possesses, when young, two eyes, a tubular case, and a single rotatory organ, which, when expanded, presents the appearance of four leaves, fringed with numerous cilia, as shown in the figure. Below this complex apparatus the mouth is situated, the jaws of which are armed with rows of teeth, which are discerned in figure 69, near the centre of the leaves; but are exhibited more highly magnified in figure 70. This animalcule has been found to be endowed with a nervous system; and two tubes are situated near the neck that apparently subserve the purpose of respiration.

The body of the Melicerta is transparent, but the enclosing cell is of a brown-

Fig. 70. Fig. 69.

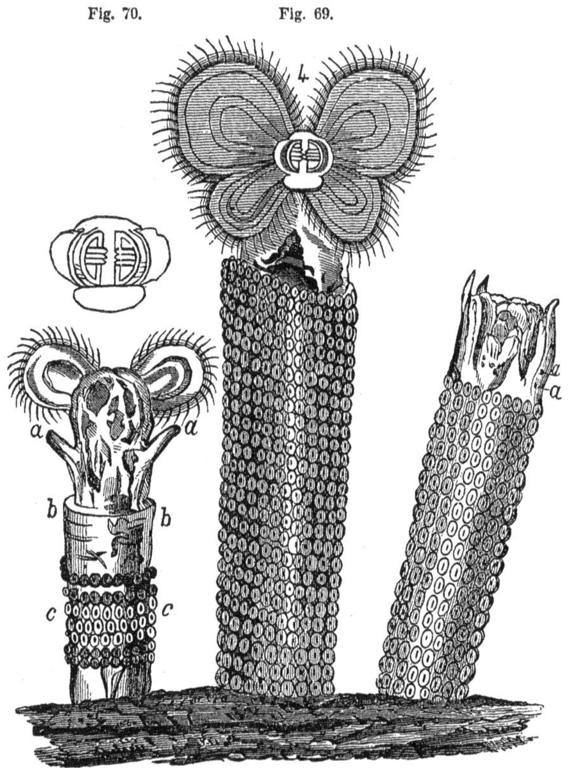

Fig. 71. Fig. 72.

ish nue, somewhat conical in shape; and its surface is composed of a numerous collection of small, regularly formed bodies, like beads; from which resemblance this species of Melicerta derives its name. These bodies are arranged in circular rows, as seen in the figures, and present a very beautiful appearance. The enclosing case in the young Melicerta is at first clear and pellucid like crystal; but as it gradually enlarges rings of beads commence forming around it, until at last the whole surface becomes entirely covered with them.

The appearance of the Melicerta, at an early stage of its existence, is exhibited in figure 71, where the tubes of respiration are shown at *a a;* the delicate transparent case at *b b;* and the first formed circles of beads at *c c.* The bead-shaped bodies are deposited by the animalcule itself, and are cemented together by a glutinous matter which exudes from its body and hardens by exposure to water.

The surrounding shell of the Melicerta is inflexible, and the soft and tender animalcule can withdraw itself at pleasure within its protecting envelope. The creature is exceedingly sensitive, and shrinks into the concealment of its case upon the slightest motion of the water in which it lives. It is seen partially withdrawn in figure 72, where the rotatory organs are seen closed up, and the two eye-specks are detected at *a a*. The Melicerta is found upon the leaves of duckweed and other water plants. Its size, when expanded, is *one-twelfth* of an inch; the length of the case *one twenty-fourth;* and that of the eggs are *one-one hundred and fiftieth of an inch.*

Fig. 73.

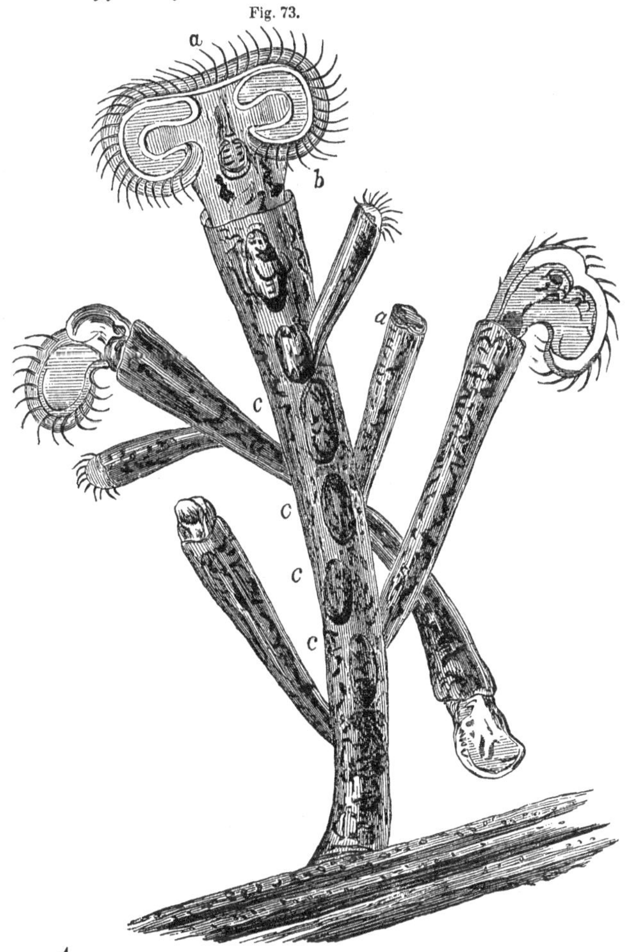

THE HORNWORT LIMNIAS, or WATER NYMPH.—This animalcule, like the one immediately preceding, is enclosed in a cylindrical case, which, at first, is white and transparent; but afterwards assumes a brownish hue. The matter com posing the case is glutinous, and extraneous particles often form a coating upon its smooth surface. Unlike the Melicerta, its rotatory organ is divided into two leaves only, fringed with vibrating cilia.

The Limnias has two red eyes, which can only be discerned when the animalcule is very young. These organs, together with the jaws, may be seen in the Limnias in the egg, before it has burst the transparent shell. In figure 73, a group of these interesting Infusoria are delineated as they appear attached to a stem of hornwort; a plant of which they are so fond that they have been designated by its name. The several individuals are here seen more or less protruded from their cases; for, like the rest of the flower-wheel animalcules, to which they belong, they are endowed with the power of extending themselves beyond the margin of their cases, and of shrinking completely within them. The parent animalcule (*a*) has its wheelwork fully protruded; its jaws and teeth are apparent at *b*, and within the sheath a row of eggs (*c c c c*) are visible In figure 74, a young Limnias is represented as it appears when just escaped from the egg; in this minute specimen the jaws and teeth, and the two red eye-specks are clearly perceived at *a* and *b*. The length of the Limnias is about *one-twentieth of an inch*, and that of the case one half the size of the animalcule.

Fig. 74.

b
a

THE ELEGANT FLOWER-SHAPED ANIMALCULE.—Another type of the flower-shaped animalcule, and which, from its beauty, has received the above name, is represented in figures 75 and 76, upon the stem of a water-plant. It is enclosed in a delicate and flexible crystalline case (*a*) and its rotary organ is divided into six leaves, (*b b*) from the ends of which brushes, formed of very long filaments, project. This creature is capable of expanding and contracting itself to a very great extent; for at one time it can thrust out nearly the whole of its body beyond its sheath, as seen in figure 75; and at another conceal itself completely within, leaving nothing but the long cilia projecting without, as displayed in figure 76. In extending itself, the flower-shaped animalcule moves slowly; but its contraction is quickly performed; and in effecting this change in its shape, the animalcule not only shortens its body, but also the flexible case, which gathers down upon itself in circular folds. They are very voracious creatures, feeding upon great numbers of monads, and the little ship-animalcules, which can often be distinctly seen within the stomach, as shown at *c*. The position of the jaws and teeth, with which they crush and tear their prey, is indicated by the letter *e*; and their structure and arrangement are apparent in figure 77, which represents this formidable apparatus very highly magnified. In figure 76, a young animalcule, with its two eyes, is seen at *f*, in the envelope containing the eggs of the parent. The size of the Flower-shaped animalcule is about the *one-one hundred and eighth part of an inch.*

Fig. 75.

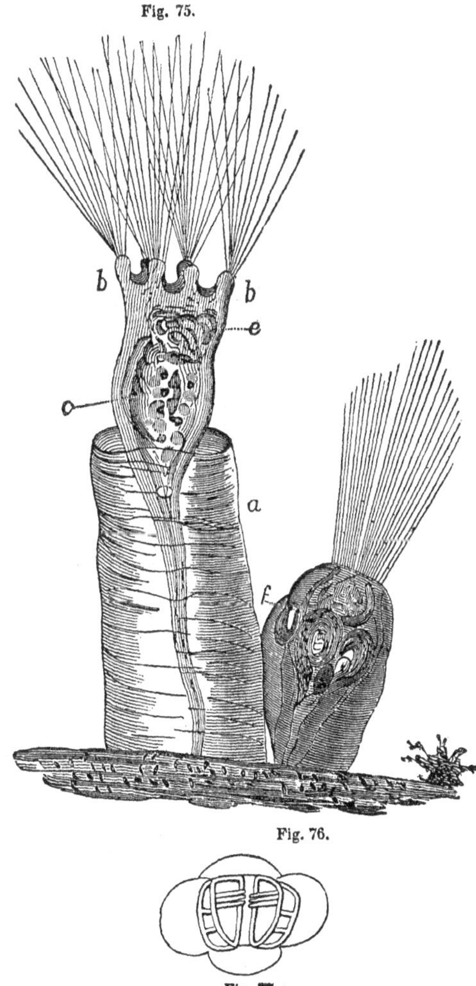

Fig. 76.

Fig. 77.

CHAPTER II.

MICROSCOPIC FOSSILS.

" All that tread
The globe, are but a handful, to the tribes
That slumber in its bosom." —*Bryant.*

" Where is the dust that has not been alive?"—*Young.*

WHEN the encased protophytes die, their soft and gelatinous parts quickly decompose; but their shells or cases remain, retaining for ages their peculiar forms and structures. To such an extent do these minute plantules swarming throughout the waters of the globe, increase, by their various modes of production ; and so rapidly do these myriad generations succeed each other, that the shells of organisms which perished centuries ago are now found in a fossil state,constituting a large proportion of the materials of extensive tracts of land, several feet in thickness, that cover the surface of the earth for many miles. These cases consist for the most part of lime, iron, and flint, and entire ranges of hills and masses of rock are composed of these minute envelopes. Dr. Ehrenberg has ascertained, that no less than five kinds of rocks and mineral substances consist wholly or in part of the fossil shells of organisms, and that three other kinds have probably the same origin. Bog iron is made up of microscopic iron shells, and the remains of the Diatomaceæ have been abundantly discovered in beds of marl. So numerous are these fossil coverings amid the chalk cliffs, that they are detected in the smallest portion of chalk that can be taken up on the point of a knife. The deposits at the mouth of rivers frequently consist, to a large extent, of protophytes, both living and fossil; and the land is thus, in many places, continually advancing upon the sea, from a cause which, until a few years ago, had entirely escaped observation.

The searching investigations of distinguished naturalists have furnished a most interesting fund of facts, which fully attest the truth of the above remarks. In Bilin, in Bohemia, a mass of slate has been discovered, forming a series of strata *fourteen feet thick*, almost entirely composed of the flinty shells of the Diatomaceæ. It is used when ground, as a polishing powder, under the name of *tripoli*. A single druggist's shop in Berlin disposes yearly of more than *twenty hundred weight*, and the supply is still sufficient for the demands of trade. The smallest amount of this powder, when examined by the microscope, is seen to be full of the fossil remains of plantules, as is likewise true of tripoli from other localities. A cubic inch of the Bilin-stone weighs two hundred and twenty grains, and contains no less than (40,000,-

000,000) *forty thousand millions* of distinct, organic forms. The species of Diatomaceæ, of which nearly the whole mass is compacted, is the divided Gallionella, or box-chain organisms; a kind of Diatomaceæ which has already been described. A specimen from this slate is delineated in figure 78, magnified three hundred times. Its natural length does not exceed one-sixth of the thickness of a human hair, and the flinty shell of a single gallionella weighs only the *one hundred and eighty-seven millionth part of a grain.* The identity of the fossil and living plantules is seen at a glance by comparing the engravings in which they are respectively represented. In Virginia, extensive beds of flinty marls have been discovered by Professor Rogers, composed, in a great measure, of the shells of different species of marine organisms. The towns of Richmond and Petersburg are built upon these strata, which vary in thickness from *twelve to twenty-five feet,* and comprise tracts and districts of considerable extent. So full is this earth of microscopic fossil remains, that when a little of it has been mixed with a drop of water, and the liquid has evaporated from the glass slide, the smallest stain left upon the surface abounds with curious vegetable structures, whose living types inhabit, to a great extent, the neighboring seas.

Fig. 78.

In figure 79 are shown two species of Navicula, which, with several others, have been recognized in the Richmond earth; but the most exquisite structure here revealed is a beautiful saucer-shaped shell, the surface of which is divided into hexagonal or six-sided figures, like the cells of honey-comb. The protophyte to which it belongs is called, from the appearance of the shell, the *Coscinodiscus,** or sieve-like disk: there are several species of these organisms, whose shells vary in size from *one-hundredth to one-thousandth of an inch in diameter.*

Fig. 79.

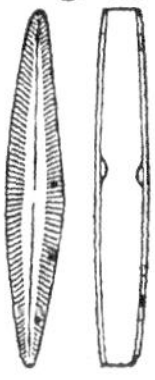

In figure 80 is displayed a portion of the circular shell of an elegant species found in the Virginia marl, which has received the name of the *Radiated coscinodiscus.* It is shown very highly magnified, and the rich and perfect arrangement of symmetrical forms here exhibited, is but a faithful copy of the wondrous original. These beautiful fossil shells are not confined to the Richmond locality, but have been discovered in the chalk marls of Zante and Oran; and Col. Fremont likewise found them in Oregon, at the Riviere Aux Chuttes. The various species of this protophyte exist in a living state in the sea near Cuxhaven, at the mouth of the Elbe; and the radiated coscinodiscus has also been detected in the waters of the Baltic, near Wismar.

Fig. 80.

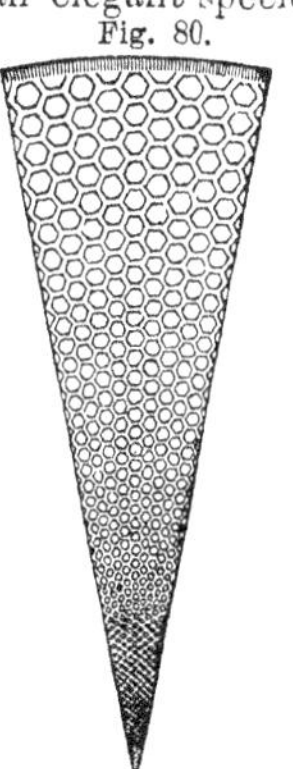

A like deposit of microscopic shells, *fifteen feet thick,* exists at Andover, Ct., and Ehrenberg remarks, in his memoir on the Microscopic life of North and South America, "that similar beds occur by the river Amazon, and in great extent from Virginia to Labrador."

* From *koskinon* (Greek), *a sieve.*

In Sweden and Lapland, a white, mealy earth is found distributed in layers, sometimes thirty feet in thickness. It is wholly composed of the shells of Diatomaceæ, and 'when mixed with the ground bark of trees is used by the inhabitants as an article of food in times of scarcity. The same kind of earth occurs in San Fiora in Tuscany, and also near Egra in Bohemia, about three feet below the surface of the ground. To the eye it appears when dry like pure magnesia; but when examined by the microscope, it is seen to consist entirely of a richly figured species of a Diatom, which is called the Campilodiscus. A specimen from this locality, very highly magnified, is delineated in figure 81. Its natural size varies from *one-four hundred and thirtieth* to *one-two hundred and fortieth of an inch.* In the province of Luneberg, in Saxony, a layer of this kind of earth also occurs, *twenty-eight feet* in thickness, which is the greatest deposit that has yet been discovered: and similar strata have been found in Africa, Asia, and the South Sea Islands. On the banks of the Amazon, in South America, a bed of fine clay occurs of the same nature. It is not a recent deposit from the swelling of the river; but is an ancient bed whose age is undetermined, and exists as an elevated and extensive plain, shaded with woods and the thick foliage of forests.

Fig. 81.

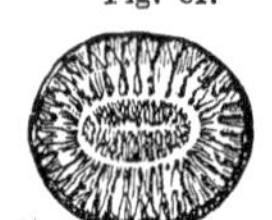

Microscopic Fossils in Chalk and Flint.—Chalk consists in a great measure of fossil structures, together with minute shells, so exceedingly small that a *million* distinct structures are computed by Ehrenberg to be contained in the space of a cubic inch. These organic remains constitute nearly half the bulk of the chalk of Northern Europe, and exceed this proportion in that of Southern Europe. The portion of these chalk formations that is not organized was originally shells, which having become decomposed, now form a cement for the organic remains, uniting them together in one compact mass. The larger shells are perceived, when the sediment obtained by brushing chalk into water is closely examined; but in order to detect the true microscopic structures, the following process must be adopted, which has been pursued by Ehrenberg. A drop of water is first placed upon a thin slip of glass, and then upon the water as much scraped chalk must be spread as will cover the fine point of a knife. After leaving the chalk to rest for a few seconds, the finest particles suspended in the water must be withdrawn, together with most of the liquid, and the remainder suffered to become perfectly dry. This sediment must now be covered with Canadian balsam, and the glass held over a spirit lamp until the balsam becomes slightly fluid without froth or air bubbles. In this state it is kept for a short time, until the balsam thoroughly penetrates every part of the sediment, flowing into the chambers and cavities of the microscopic shells, and causing their structure to be more readily detected. When a preparation thus made is magnified three hundred times the chalk is seen teeming with minute organic forms, the peculiarities of which are so clearly revealed, that the observer is enabled to arrange and classify them with the utmost ease. Flint, to a large extent, has also been proved to be of animal origin; and a distinguished English naturalist has observed, that masses of flint, or nodules as they are termed, are *almost entirely composed* of the shells of minute animals, mingled with the scales of fishes, zoophytes, and the remains of numerous minute animals. The microscopic animal structures that abound most in the chalk and flint of England are two kinds of Polythalamia,* or many chambered shells; termed the Rotalia,† or *wheel-shaped* animal, and the curious Textularia,‡ or *entwined* animal. With these are combined vast numbers of minute shells, belonging to an extensive class of small animals, which, on account of their being covered with pores, have received the name of Foraminifera.§

The shells of the Foraminifera differ in their dimensions. Some of them are perfectly microscopic, being invisible to the naked eye; while others are of the size and shape of a dollar; and from their resemblance to a coin have received the name of Nummulites,‖ or *fossil-money*.

Fig. 82. Fig. 83. Fig. 84.

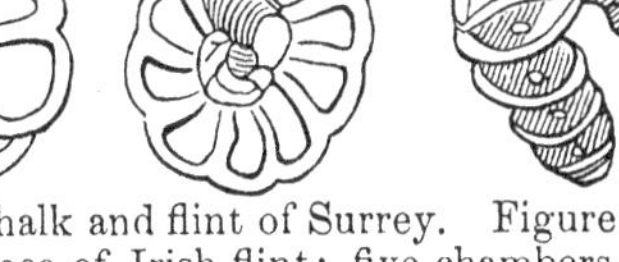

In figures 82 and 83 are delineated two microscopic shells of the Rotalia, each of which is seen to consist of several compartments, like that of the nautilus; though they are distinct from the latter in their nature. The specimens, from which the original drawings were taken, were discovered in the chalk and flint of Surrey. Figure 84 represents a portion of a nautilus found in a piece of Irish flint; five chambers of the shell are clearly seen, partially separated from each other. The three figures here presented are all very highly magnified.

A beautiful species of microscopic fossil, that is likewise found in chalk, is the *Crosier-like* shell, which in its advanced state changes its original shape, and assumes the graceful form shown in figure 85, which presents a side view of the object. This fossil was found at Chichester, by Mr. Walter Mantell, and is here shown as it appeared when magnified eight times.

Fig. 85. Fig. 86. Fig. 87.

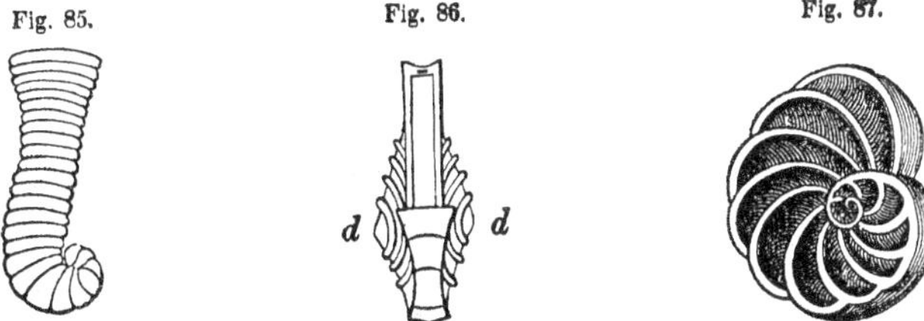

Another kind of the microscopic many-chambered shells is the *Fan-shaped* animal fossil, which occurs abundantly in the chalks of France, and is also found in those

* From *polus* (Greek), *many*, and *thalamos* (Latin), a *chamber*.
† From *rota* (Latin), a *wheel*.
‡ From *textura* (Latin), *woven-work*.
§ From *foramen* (Latin), *opening*, and *ferre* (Latin), to *bear*.
‖ From *nummus* (Latin), *a coin*, and *lithos* (Greek), *a stone*.

of England. A profile of this shell, magnified twelve times, and bearing some resemblance to a fan, is shown in figure 86. When a side view is taken, and the fossil is highly magnified, the beauty of the structure becomes more apparent, and the fluted projections, *d d*, are revealed as elegant spiral shells, divided into several apartments, and presenting an appearance similar to that which is exhibited in figure 87.

Fig. 88.

Fig. 89.

The *Textularia* or *entwined* animal has the figure of a cluster of globes, rising in the form of a pyramid, and when a section is made in the direction of its length, it displays the different cells into which the cavity of the shell is divided. In figure 88 is shown a specimen from the marl of the Mount of Olives, and an outline of the American sextularia is also exhibited in figure 89. This species differs in some respects from other Textularia, being wholly local and peculiar to the chalk marls of the Upper Missouri; of which vast deposit it forms the principal part. The living Xanthidia, or Cross-bar protophytes, have already been described; and in figures 90, 91, 92 and 93 are presented several specimens as they appear in flint. In this stone they often occur in great abundance, no less than twenty being once discovered by Mr. Hamlin Lee, in a chip of flint, the surface of which was scarcely the *twelfth* of an inch in diameter. These organisms are easily detected in flints which are *translucent;* the only preparation required being simply to select the thinnest and clearest flakes, struck off by the blow of a hammer, and before viewing them with a microscope, to immerse them in oil of turpentine, in order to render them more transparent. The specimens of Xanthidia represented in figures 90, 91, 92 and 93, were taken from a remarkable group, described by Dr. Mantell, and found by his son in a

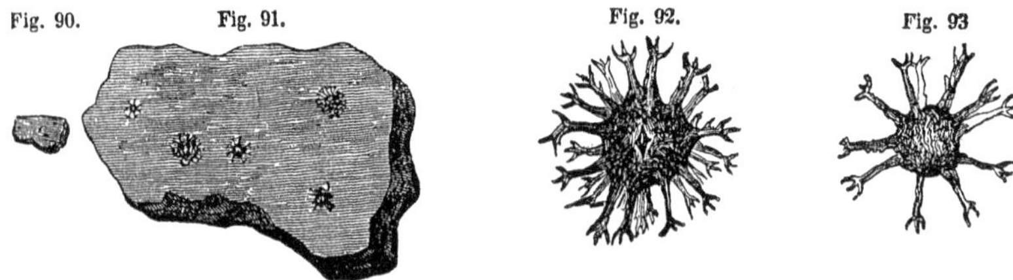
Fig. 90. Fig. 91. Fig. 92. Fig. 93

flake of flint. This flake is delineated of its natural size in figure 90; in figure 91 it is considerably magnified, and the several fossils are distinctly seen. Figures 92 and 93 are two of the specimens very highly magnified, and are a variety of the *Branched Xanthidium*, which is found only in a fossil state. That they belong to the race of the Xanthidia is evident from the resemblance they bear to the drawings of the living specimens, figures 40 and 41. Five specimens were found in this fragment of flint, varying in diameter from *one-three hundredth* to *one-five hundredth* of an inch.

PEAT BOGS.—The peat bogs both of ancient and modern origin, are frequently found to contain beds and layers of a white flinty earth, which is entirely composed of the shells of protophytes. In many swamps of Ireland and England, earthy strata of this peculiar nature have been found; and in this country, Prof. Bailey has discovered near West Point a deposit eight or ten inches thick, and in all probability several hundred yards in extent, wholly made up of the flinty shells of the diatomaceous organisms, in a fossil state. "This deposit," says Prof. B., "is about a foot below the surface of a small peat bog, immediately at the foot of the southern escarpement of the hill on which the celebrated Fort Putnam stands. In draining this bog a large ditch was dug, and among the matter thrown out, my attention was attracted by a very light white or clay colored substance, which, when examined closely in the sunshine, showed minute, glimmering, linear particles. On submitting it to observation, by means of a good microscope, I found it to be almost entirely composed of fossil organisms. There can be no doubt, that in this place there are *several tons* of the shells of beings so minute as to be barely visible as brilliant specks, when carefully observed in a strong light by the naked eye. Hundreds of years must have elapsed before such an accumulation could have been made." The kind of shell that is most abundant in this earth is delineated in figure 94, which represents a specimen magnified three hundred and fifty times; and in figure 95 is shown the appearance presented by a little of the earth diffused in a drop of water, and magnified about *fifty times*. The earth is here seen consisting of a great number of shells of various shapes and sizes, clearly proving that the deposit is nothing more than a vast assemblage of immense multitudes of minute fossil structures.

Fig. 95.

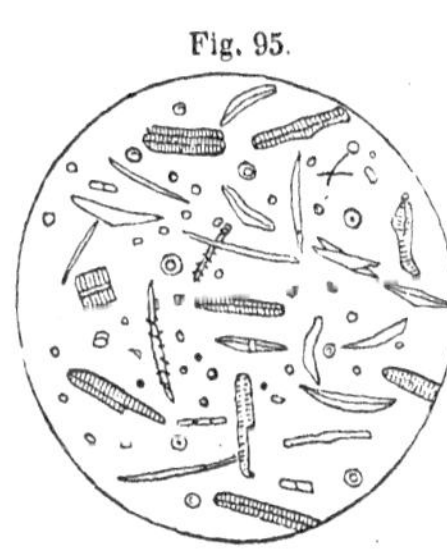

Fig. 94

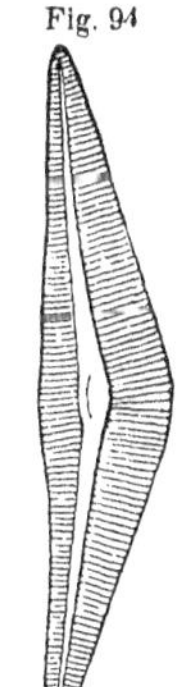

FORAMINIFERA.—The fossil shells of these minute forms of animal life now exist in such profusion, rising into mountains, and extending in broad and deep layers beneath the surface of the earth, that it has been observed by the learned Dr. Buckland, "that the remains of such minute animals have added much more to the mass of materials which compose the exterior crust of the globe, than the bone of elephants, hippopotami, and whales."

In these vast collections the *Nummulites* largely prevail. They are divided into numerous species, varying in dimensions from the size of a crown-piece to that of a grain of sand. The spiral shell of the nummulite is delineated in figure 96: it is separated into a very great number of small cells of nearly equal extent, which communicate with each other by an opening through the partitions of the several chambers. It is supposed that each cell once contained a distinct animal, and that the entire shell formed the common habitation of a vast multitude. The chalk formation at Bayonne and of the Pyrenees, consists of beds of crystalline marble, composed of nummulites, and the vast limestone range at the head of the Adriatic Gulf, is also constituted of nummulites, having the shape and size of a small pea. At Suggsville, in the United States, is a chain of mountains three hundred feet high, entirely made up of a single species of this fossil. The great pyramid of Egypt, which covers eleven acres of ground, and rises to the height of about 600 feet, is constructed partly of limestone, which consists of nummulites and microscopic fossil structures that form a cement for the larger shells.

Fig. 96.

There exists in the north of France an extensive tract of country, one hundred and eighty miles long, and about ninety in breadth, within whose limits Paris is included. This region is termed by geologists the *Paris Basin*, and the exterior crust of the earth is here composed of layers or strata of sand, marl, and limestone—alternating with beds of plaster of Paris (gypsum) and flinty matter. These vast beds of marl and limestone are full of foraminiferous and other forms, and deposits of great thicknesses have been discovered, which are entirely constituted of nummulites no larger than a grain of millet seed. The limestone from the quarries of Gentilly abound to such an extent with microscopic structures, that a cubic inch is calculated to contain on an average no less than 58,000 shells, and the beds thus constituted are of great extent and thickness. It is even asserted by geologists as an undoubted fact, that the edifices of the splendid capital of France, as well as of the towns and villages of the neighboring provinces, are almost entirely built of stones composed of the shells of foraminiferous animals; and that these minute fossils are scarcely less numerous in other tertiary formations, extending in the south of France from Champagne to the sea. They likewise abound in the strata of the Gironde, and in those of the basin of Vienna. The invisible, calcareous polythalamia, or many-chambered shells, form, according to Ehrenberg, the compact earth and rocks of Central North America, and constitute immense deposits at the sources of the Mississippi. Even the stupendous chain of the Andes, belonging, as it does, to the chalk formation, is conjectured to have been originally composed of minute organized remains, which have since been changed by volcanic action.

Vast beds of microscopic forms occur in Patagonia, the extent and arrangement of which is thus described by Darwin: "Here along the coast for hundreds of miles, we have one great tertiary formation, including many tertiary shells, all apparently extinct. The most common shell is a massive, gigantic oyster, sometimes even a foot in diameter. The beds composing this formation are covered by others of a peculiar, soft, white stone, including much gypsum, and resembling chalk, but really of the nature of pumice stone. It is highly remarkable from its being composed, to at least *one-tenth* of its bulk, of minute fossils, and Prof. Ehrenberg has already recognized in it *thirty marine forms.* This bed, which extends for five hundred miles along the coast, and probably runs to a considerably greater distance, is *more than eight hundred feet* in thickness at Port St. Julian." In volcanic products, which have been necessarily subjected to the action of the most intense heat, the remains of protophytes have been detected, incredible as it may appear. The Island of Ascension is of volcanic origin, and portions of a pink-colored, porous rock, which had once been flowing lava, were here taken and preserved by Darwin. These specimens were examined by Ehrenberg, who discovered, among other ingredients of which they were composed, the flinty shells of *fresh water* protophytes.

A large part of the sand of the great African desert consists of the fossil shells of small animals; and such is the fact in regard to the valley of the Nile. Numerous specimens of the deposits of this river, taken from various localities along its course from Nubia to the Delta, have been carefully examined by Ehrenberg; and in such profusion were fossil sponges, the flinty cases of Diatoms, and various species of Polythalamia discovered, that not a particle of this soil of the size of half a pin's head could be found, in which (allowance being made for certain chemical changes that had occurred) there was not one, and often several, of these fossil organisms.

MUD-BANKS.—In the harbor of Wismar, on the Baltic, there is deposited, every year, as appears from official documents, 228,854 cubic feet of mud; and the accumulation has continued at this rate for more than a *hundred years*. In the course of a century a deposit has therefore been made to the extent of 22,885,400 cubic feet, equal to 3,240,000 hundred weight. These mud-banks were examined by Ehrenberg in 1839 and 1840, and the surprising discovery was then made, that from *one-twentieth* to *one-fourth* of the sediment was composed partly of living Diatoms, and partly of the flinty shells of others that had perished. On an average *one-tenth* part of the entire mass consists of microscopic forms, and hence the annual deposit of these structures in the port of Wismar amounts in bulk to 22,885 cubic feet, which, if it was dried, would weigh not far from *forty tons*. In the mud-banks of Pillau, the remains of Diatoms were found in greater abundance than in those of Wismar. At both localities many of the forms were entirely new, and others were identical with living protophytes that inhabited the waters of the neighboring seas.

The mud deposited by the Elbe at Cuxhaven, was found by Dr. Ehrenberg to be extremely rich in organic remains—nearly *half* of the sediment consisting of the flinty cases of Diatoms, and various species of the Polythalamia or many-chambered shells. The flinty cases of protophytes have been found at the bottom of the ocean in the mud of the coral islands beneath the equator, and no less than *sixty-eight* species have been discovered in the mud at Erebus Bay, near the Antartic pole. The examination of the sediment deposited along the Atlantic coast of America, has revealed similar facts. Diatomaceous forms have been detected in the mud of Boston harbor, and in the marine marshes at New Haven in Connecticut; and numerous elegant infusorial structures and many-chambered shells have been found at Amboy in New Jersey, in the mud adhering to oysters as they were taken from their beds.

In view of facts like these, it has been asserted by naturalists, that the deposits in harbors, and the accumulation and amazing fertility of the mud of the Nile, and probably of other turbid rivers, are to be attributed in a great measure to the agency of invisible forms, whose countless generations succeed each other with astonishing rapidity, leaving the curious structures in which they were encased as the durable records of their existence.

These gradual accretions have been accumulating for centuries, and are at this moment still in progress. The sea now swarms with races of minute organisms, whose fossil types are continually discovered in beds and strata of unknown antiquity. In salt water, taken from Cuxhaven and various other places, no less than *twenty genera* and *forty living species* have been discovered by Ehrenberg, which he regards as identical with those occurring in the chalk formations. And out of *twenty-eight species* of fossil structures belonging to the various species of the Diatomaceæ, he has detected *fourteen* fresh water and five marine species, now living; the remaining nine are either unknown or extinct forms.

The Diatomaceæ that crowd the seas are devoured in multitudes by the common scallop and other molluscous animals; for when their stomachs are examined they are found to contain thousands of microscopic flinty shells, which, from their nature, were incapable of being digested. When a few atoms of the food which a scallop has taken into its stomach is viewed by the microscope, it is found teeming with a rich collection of curious shells, closely resembling the beautiful structures that constitute the Richmond deposit, not only in form but in arrangement—so striking is this resemblance, that it is said to be extremely difficult to distinguish between the recent and ancient remains; and that even an experienced observer would be liable to confound them, unless the glass slides, upon which they were mounted, were labeled.

The guano imported from the isle of Ichaboe has been found to contain the beautiful shell of the *Coscinodiscus*, and other Diatomaceous structures of great elegance and richness; and, as we gaze upon these minute cases, we cannot fail of being struck with the fact of the great resistance to decomposition which they possess. In this instance they must have gone through the process of digestion twice, and been subjected to the action of the elements for centuries. Guano, as is well known, is found within certain latitudes on uninhabited islands, which have been, for ages, the abode of innumerable multitudes of marine-birds. It consists of their excrements, which have been accumulating for century after century, until they form layers of great thickness; many beds having been discovered in the islands of the Pacific, off the Peruvian coast, having a depth of thirty-five or forty feet. The siliceous shells, detected in the guano, are the remains of Diatoms devoured by fish, which, afterwards, became the prey of voracious sea-birds. Thus the shell passed through the stomach twice, and then remained in the guano-bed for an unknown length of time, subjected to those common causes of decay which turn the solid rock itself to dust. But under all these influences they continue unchanged, and the eye of the naturalist at last detects these minute structures still possessing their original beauty, with the delicate tracery of their rich configurations, almost as sharp and clear as it was, perhaps, a thousand years ago.

DUST SHOWERS.—The fossil shells of protophytes, which lie mingled with the soil of the earth, are not unfrequently carried up into the air in the clouds of dust that are raised aloft by the winds, and borne along on the currents of the atmosphere, to a distance almost incredible. Darwin noticed that the atmosphere of St. Jago, one of the Cape de Verde isles, is generally hazy, owing to the fall of an impalpable fine dust, of a brown color. A small quantity of the dust was collected by this gentleman, who received also, from Mr. Lyell, four packets of the same kind of powder, which fell on a vessel a few hundred miles northward of the Cape Verde Islands. Five parcels were sent to Dr. Ehrenberg for examination, who found it to consist chiefly of the flinty cases of Diatoms, and the siliceous tissue of plants. No less than *sixty-seven* distinct kinds of rare structures were detected of which sixty-four were fresh-water species, and the remaining two marine.

The same observer, upon investigation, met with fifteen accounts of dust falling upon vessels when far out on the Atlantic, off the coast of Africa. It has been here known to descend upon the decks of ships, at the distance of *several hundred* and even a *thousand* miles from shore, and when land was distant to the north and south, full *sixteen hundred miles.* The dust is distributed thickly through the air, soiling everything on board, injuring the eyes, and rendering the atmosphere so hazy that vessels have been known to run ashore in consequence of the obscurity thus produced. This dust is believed to come from the African continent, from the fact that it occurs when the wind is from that direction, and at the same time that the *harmattan* prevails, which is a periodical wind that blows from the interior of Africa towards the Atlantic. Clouds of the finer particles of sand, from the arid deserts of this continent, are borne aloft by the sweep of the harmattan, and carried far out over the sea upon the higher currents of the atmosphere. At the distance of three hundred miles from land, Darwin discovered, in the fallen dust, particles of stone the *thousandth part* of an *inch square*, mixed with matter still finer.

In reflecting upon the facts just adduced, we see, that in order to become acquainted with the structure of the world we inhabit, it is not sufficient to trust to our unassisted vision. Wonders, and problems the most curious and interesting, will meet the gaze of the naturalist at every step he takes; but unless he explores the secrets of nature with the magic glass of the microscope, half of the treasures of truth will be still unrevealed: sealed from his vision by impenetrable darkness.

A question naturally arises, What are the ends which these microscopic animal organisms, whose modes of existence, habits, and structures have been discussed in the preceding pages, subserve in the economy of the world? To them have been attributed malignant influences; for the various epidemics, which at intervals have swept down our race, have been supposed by some to originate in a "living cloud" of existences, dwelling in the air. But of this we know nothing certain, and a more satisfactory answer cannot be given than that which is contained in the words of Professor Owen, who thus unfolded his views upon this subject, in one of his lectures:—"Consider their incredible numbers, their universal distribution, their insatiable voracity, and that it is the particles of decaying vegetable and animal bodies which they are appointed to devour and assimilate. Surely we must, in some degree, be indebted to these ever active, invisible scavengers for the salubrity of the atmosphere, and the purity of water. Nor is this all; they perform a still more important office in preventing the gradual diminution of the present amount of organized matter upon the earth. For when this matter is dissolved or suspended in water, in that state of comminution and decay which immediately precedes its final decomposition into the elementary gases, and its consequent return from the organic to the inorganic world; these wakeful members of nature's invisible police are everywhere ready to arrest the fugitive organized particles, and turn them back into the ascending stream of animal life. Having converted the dead and decomposing particles into their own living tissues, they themselves become the food of larger Infusoria, and of numerous other small animals, which, in their turn, are devoured by larger animals: and thus a food, fit for the nourishment of the highest organized beings, is brought back by a short route from the extremity of the realms of organized matter. These invisible animalcules may be compared, in the great organic world, to the minute capillaries in the microcosm of the animal body; receiving organic matter in its state of minutest subdivision, and when in full career to escape from the organic system, and turning it back by a new route towards the central and highest point of that system."

CHAPTER III.

MINUTE AQUATIC ANIMALS.

"Then sweet to muse upon His skill displayed
(Infinite skill) in all that he has made!
To trace in nature's most minute design,
The signature and stamp of power divine:
Contrivance intricate, expressed with ease,
Where unassisted sight no beauty sees;
The shapely limb and lubricated joint
Within the small dimensions of a point;
Muscle and nerve miraculously spun,
His mighty work who speaks and it is done."—COWPER.

THE POLYPE.—This singular animal, of which there are several species, is found abundantly in ponds and brooks, attached to the leaves of aquatic plants, and to the surface of twigs and branches that have fallen in the water. Its body, which simply consists of a collection of cells, formed of grains of green and brown matter, possesses the power of expansion and contraction, and appears, when extended, in the shape of a jelly-like tube, about the size of a bristle; tapering from the upper to the lower extremity, and having a length ranging from one-quarter to three-quarters of an inch. The mouth is furnished with feelers or arms, which vary in number, in different specimens, from six to sixteen, and are employed by the animal for the purpose of seizing its food. Though appearing to the unaided eye as attenuated threads, the microscope shows them to be, in fact, slender tubes filled with a fluid, and consisting of a series of cells like the body of the animal. When contracted the polype appears like a tiny ball of jelly, hardly one-tenth of an inch in diameter, and the long arms or feelers shrink into little conical eminences, ranged in a circle around the upper part of the body.

Figures 97, 98, 99 and 100 present a magnified view of several polypes, in different states of contraction, with their prey within them. A species, termed from its color the Green polype, is delineated in figures 97 and 98 and another kind, the Brown polype, is represented in different attitudes in figures 99 and 100. The small circles exhibit the specimens of their natural size. The mouth of the polype is unfurnished with teeth, and presents different appearances, according as it is more or less contracted; at one time assuming the form of a cone, and at another appearing cup-shaped, with an aperture in the centre, capable of great expansion for the reception of its food. This last form is shown in figure 98, where the animal is seen gorging its prey. The polype feeds upon small crustaceous animals, worms and larvæ;* and when in search of food extends its body and

* Larvæ are the young of insects in their caterpillar state.

Fig. 97.

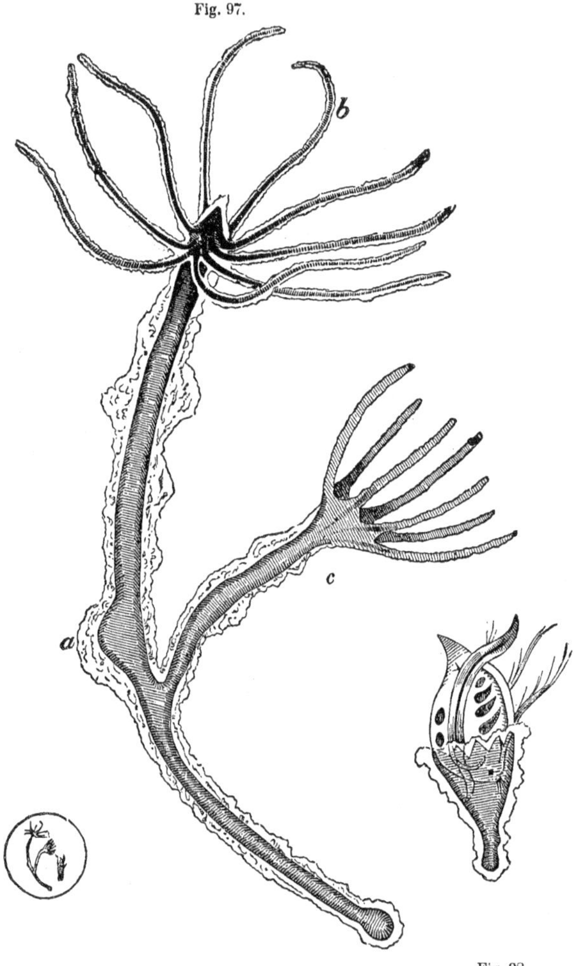

Fig. 98.

feelers to the utmost, spreading out the latter in different directions, so as to command an extensive field; as soon as an animal comes within their range the feelers twine themselves around it, and gradually contracting, convey the prey to the mouth of the polype. A polype in the attitude of watching for its prey is represented at *b*, in figure 97. It sometimes occurs, that the animal attacked

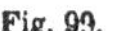

Fig. 99.

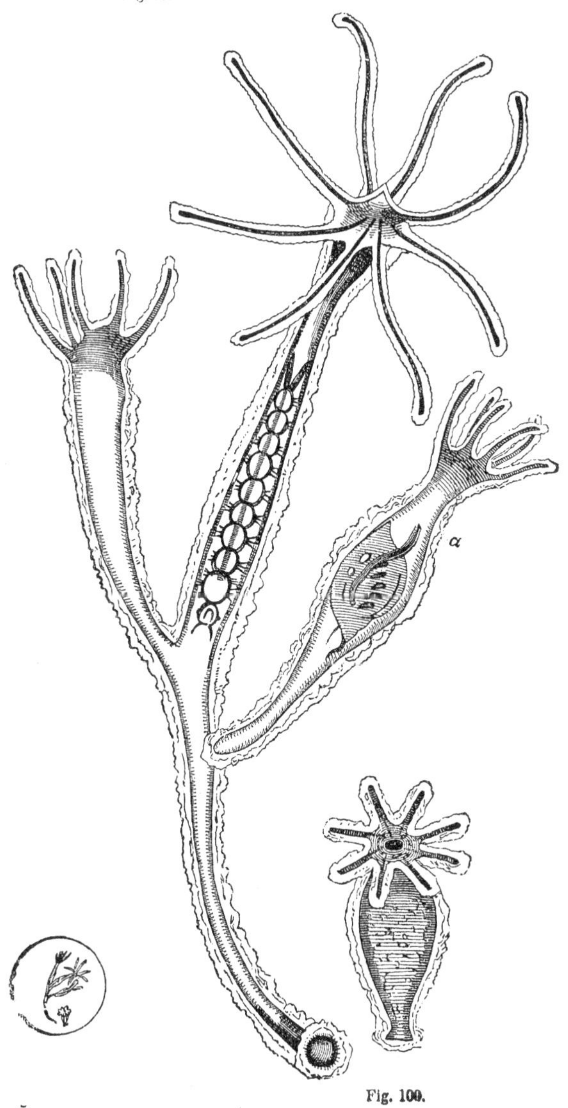

Fig. 100.

by the polype moves so rapidly as to prevent the assailant from instantly securing his victim; in such a case the latter is seen to sink into the water after its attack, and remain to all appearance lifeless for the space of a few seconds, before it regains its usual vigor. Naturalists have consequently been led to suppose that the polype possesses the power of paralyzing its prey by weak electric shocks, in the manner of the torpedo and the electric eel. In this way only can they account for the fact, that such slender organs as the arms of the polype are able to secure animals comparatively so large and powerful, when striving with their utmost power to escape from the fatal coils in which they are entwined. Dr. Mantell once beheld a lively polype seize two large worms at the same instant, when its extended arms were so attenuated that they were scarcely visible without the aid of a lens; and yet the worms, though struggling desperately for their lives, were unable to burst from the slender bonds that encircled them, and in an instant lost all power of motion: the same effect is produced upon the Water-flea, an extremely vivacious little creature, when struck by the feelers of the polype. The power exerted by the arms is considered to be electric in its nature, inasmuch as the polype has never been found to possess a sting or destructive weapon of any kind.

The stomach of the polype consists of the whole internal cavity of the creature, and when its prey has been seized and devoured, the body and feelers are no longer extended, but contract, as shown in figures 99 and 100, where *a*, figure 99, represents a polype partially contracted, and figure 100 one entirely so. While the process of digestion is advancing, the polype is very sluggish, and the whole nutritive fluid is disseminated throughout the internal surface, both of the body and the feelers, imparting to them a colored appearance; thus, when a red worm has been devoured, the hue of the prey tinges the entire surface of the polype. The polype multiplies by buds and shoots, which spring out of the trunk of the parent, as shown in figure 99. If it is kept in a vessel of water, and well provided with food, two or three shoots are seen, when the weather is warm, growing out of its body at the same time, and from these branches while yet attached to the parent trunk, other sprouts and offsets push vigorously forth. When a young polype is about to come into existence, that part of the body from which it will grow swells beyond its natural size, as shown at *a*, figure 97. This protuberance continues gradually to increase, and when a sufficient enlargement is attained the head of the young polype appears, and its arms are protruded, and by the aid of the latter it now supplies itself with food, in the manner of the parent, as seen at *c* in the same figure. Until nearly the time when it separates from its parent, the young polype possesses an internal communication with the latter, and also a common sensation; for if one is disturbed and contracts the other directly does the same.

The polype is endowed with the wonderful property of *reproducing any organs of which it has been deprived;* for its body, however mutilated, soon supplies its deficient members, and the creature becomes once more a perfect and complete animal. If a polype is divided *across* into two parts, the *upper portion*, containing the arms, speedily provides itself with a *new body and tail*, and the *lower part* pushes forth a fresh *body and head* with its slender arms. If the animal *is slit down from the head to the tail*, but is not quite severed, each of the two parts, thus left hanging together, becomes a perfect polype, and they live and roam through the water indissolubly linked to one another. Nay, more, if a polype is *turned inside out*, it soon accommodates itself to this new arrangement; for the original outer skin, now lining the interior cavity, performs the office of digestion; while the coating of the former stomach becomes the covering or skin of the polype.

The possession of this strange faculty by the polype is not a matter of inference or conjecture; inasmuch as it has been proved by experiments, beyond the possibility of a doubt. It was first discovered about a century ago, by Mr. Trembly, of Holland, whose statements were afterwards verified in England in every important particular, by the experiments of Mr. Henry Baker, of the Royal Society; and still later by Pritchard, who thus details one of his experiments. "Having selected a brown polype out of a glass vase containing a good supply of them, none of which had more than seven arms, I severed it obliquely, the upper part comprising the greater portion of the head and four arms; the lower part being the tail with the remainder of the head and two arms. These pieces were then put in a four-ounce phial of water, with a few small crustacea, where they sunk to the bottom, apparently lifeless. *Three hours* after the operation I examined them, and found them in the same state.

Twelve hours after this I found the lower part attached to the side of the phial by its tail, with its arms extended in quest of food; the upper one still remaining at the bottom, but with its arms extended like the other.

On the *second day*, a new tail was completed to the upper part of the polype, and the rudiments of additional arms were developed in both, and each portion appeared in good health. On the *third day*, the new arms were nearly of the same size as the others, and in *less than a week* each of the two polypes had a young one sprouting from it. The most curious circumstance connected with this experiment was, that the two new polypes had each *ten* arms, while that from which they were produced, as well as those which were in the same vessel, had only *six* or *seven*."

The number of parts into which this creature is divided presents no obstacle to the operation of this extraordinary law of vitality; for a single polype has been cut into *ten* pieces, and each part soon became a complete animal.

The Round Lynceus, or Monoculus.—This name is given to the curious little creature, which is shown highly magnified in figure 101. It is covered with a delicate shell, presenting by its fine reticulations the appearance of mosaic work. This envelope, with its minute divisions, is beheld in the drawing at *a*, and encases nearly the entire body of the animal. In some species the shell is adorned with diamond-shaped figures, in others its surface is composed of hexagons like that of a honey-comb; and a diversity of other angular figures embellish the cases in the different remaining varieties of the Lynceus. The shell is perfectly transparent, and consists of a single piece without hinge or joint; being sufficiently elastic to permit the animal to open it at pleasure. The position of the edges of the opening, is indicated in the figure, on the under side, by the letter *b*. Not only is the animal itself protected by this delicate case, but it affords a secure retreat for the young when danger is near. They then escape from the approaching peril by swimming within the shell of the parent, which the latter opens for their reception, and closes as soon as their entrance is effected.

The two eyes of the Lynceus at *d* are of different sizes and are of a deep black hue; while the rest of the animal is buff, approaching to orange The beak is seen at *c*, and the two horns or feelers at *g*. Within the shell is a row of four false feet, easily discerned, that assist the Lynceus in creeping along the stalks of plants, to which it attaches itself by pressing their sides with the edges of its shell in the manner of a pair of pincers. These members also subserve another purpose, causing the animal, as it advances through the water, to proceed with a revolving motion; in which action it is also aided by the appendage *l*, which, striking against the water like a fin, renders the rotatory motion of the Lynceus more rapid. This organ, which resembles a tail, is armed with two strong claws, is forked at the extremity, and fringed along the edges with rows of hairs. The Lynceus feeds on animalcules, and is the food of larger water insects. The position of the stomach is indicated in the drawing by the curved figure within the shell.

The Small Water-Flea.—This little animal is found abundantly in ponds and brooks during the summer months, sporting about in the waters with great activity. According to Pritchard, they are usually colorless in ponds covered with herbage, but in small collections of rain water in a loamy soil they glow with a fine bright red hue. A drawing of this animal, of its real size, is seen in figure 102, and a magnified representation in figure 103. The body of the Water-flea is covered with a kind of armor, formed of plates of shell that overlap each other, and are capable of being moved sideways as well as up and down.

Their ends do not meet on the underside, thus affording a sufficient space for the insertion and motion of the organs of respiration, which are seen at *a*, but are exhibited with greater distinctness at *c*, where they are more highly magnified.

The eye of the Water-flea, shown at *d*, is of a dark crimson hue, and from each side of it, spring two pairs of horns, which consist of numerous joints, studded with bristles, two or more proceeding from each joint.

In some species the sexes are distinguished by them, the males having a bulb about the middle of the right antennæ, or horns, as shown in figure 104. The appendages which are seen attached to the lower extremity of the animal are the bags containing its eggs, and which are together, nearly equal in size to the bulk of the insect itself. Below these sacks the tail is forked and adorned with a plume of fringed hair. In most instances the shell of the Water-flea is transparent like crystal, but it is frequently embellished with beautiful tints. Some are of a bluish green, and others red, with the receptacles of the eggs of a green color. In the specimen from which the drawing was taken the shell was richly adorned with bright red hues.

THE VAULTER.—The appellation of Vaulter is given to the minute insect which is represented, highly magnified, in figure 105. It derives its name from the circumstance that it transports itself from place to place by successive leaps, in the manner of a flea. As a person approaches, it remains quiet for a short time upon the leaves of the plant on which it happens to be; but soon springs away to some other place, a motion which it effects by bending its body and darting away from point to point, by the force of the recoil. In England the Vaulter appears in the greatest numbers in the months of April and May, swarming upon the stalks and under side of the leaves of healthy duck-weed, growing on the surface of the water. Stagnant water, filled with decayed plants, is destructive to them, and in order to preserve them, they must be provided with plenty of clear, pure water. The Vaulter is very active, and when caught, is usually detected in the eager pursuit of its prey. The encasing shell of this creature is similar to that of the Small Water-flea, but differs in having a greater number of parts. The body tapers also more gradually, and the horns do not contain so many joints as those of the former. It is also distinguished from this insect by having under the beak a single organ of respiration, which is delineated in figure 106. This instrument is in constant motion, and causes a current of water to set towards the animal, like the cilia of the Infusoria. The legs are ten in number, and are fringed with hairs; and the tail of the Vaulter, which consists of two parts, is ornamented in the same manner. The eye is deeply imbedded within the shell, and its position in the figure is indicated by the letter *c*. The upper part of the Vaulter gleams with a bright red hue of various tints, fading down to salmon color on the under side of the body and legs; while the tail, the tufts of hair which fringe the legs, and the horns or feelers, are of a bluish green. The Vaulter is only *one three-hundredth of an inch* in length.

THE LARVA OF A SMALL BOAT-FLY.—This insect is so called from its resemblance in form to a boat. They swim on their backs, and propel themselves with considerable force by means of their hinder feet, which are shaped like oars. Pritchard observes, that they are found during the spring, playing upon the surface of ponds and streams, and immediately seeking the bottom when disturbed. They acquire their full richness of color, and attain their perfect state in the fall months, at which time they deposit their eggs, which are small in size, and consist of a jelly-like substance. As the young advance towards maturity, they shed their skins several times; at which periods they are colorless throughout, except the eyes, which are of a light crimson.

They acquire their peculiar tints sometime afterwards, the lower part of the body changing through every variety of hue, from a pale yellow to a rich carmine. The insect possesses the form depicted in figure 107, which presents a magnified view of its under side. The head (*a*), is of a yellow color, and on each side of it are placed two reticulated eyes (*b b*), of a deep crimson hue. This creature has three pairs of feet, fringed with hair; the first of which often escape observation from their color and position. The second pair are generally thrown forward in a swimming position, as displayed in the figure. The hinder pair are the strongest, and are armed at their extremities with claws. The beak is hard and pointed, and in some of the larger species is capable of inflicting a severe puncture. In the preceding figure this organ is shown foreshortened, but in figure 108, the head and beak of the insect are represented, highly magnified. The compound eyes are seen at *d d*, and the beak at *b*. The latter consists of several parts, is cased in a horny substance, grooved down the middle, and terminates in a fine, hard point. The inner wings of the Boat-fly, when arrived at maturity, are fragile and delicate, and are protected from injury, like those of many other insects, by hard, shelly cases, under which they are neatly and compactly folded. The insect is ornamented with long hairs, which, along the lower part of the body and down the middle, are arranged in thick tufts. These tufts appear to be intended for the purpose of permitting the creature to float at pleasure, without any exertion, and this result is effected in the following way: the Larva first rises to the top of the water, and then elevating the lower end of its body above the surface, lifts up the rows of hair on either side of its body, suffering the air to fill the channel or groove that they before occupied. Retaining the air in this cavity, it thus becomes specifically lighter than the water, and now floats at its ease upon the surface. When the insect wishes to descend, it smooths down the rows of hair into their former place, by the aid of its feet, and thus expelling the air, renders itself heavier than the water, and sinks. The margin of the body of this little creature is naturally of a bright carmine, the central portions of a yellowish brown, and the legs of a delicate straw color. Its food consists of the eggs and larvæ of water-insects, and it often awaits its prey by lying at the bottom of the water with its beak upwards, ready for assault. In this position it remains, until its victim, unwarily descending, falls in a moment into the power of its destroyer.

Fig. 108.

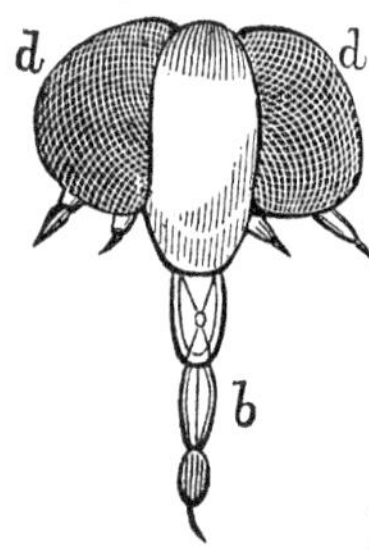

d d b

THE LARVA OF A SPECIES OF WATER-BEETLE.—The eggs from which this Larva is produced are found, in the spring and summer, adhering to the surfaces of aquatic plants. They are enclosed in a bag, a little smaller in size than a pea, which is fastened by a slight thread to the herbage.

These eggs are readily hatched by placing them in a vessel of water, and exposing them to the sun for a few days; when the young soon appear, moving about the fluid with great activity. At first they are of a dark hue, but in a short time they shed their skin, during which process their color fades; at the same time their vivacity forsakes them and they abstain from food.

As they recover, their color changes again, and diversified tints adorn their bodies. These creatures are extremely voracious; for if they are placed in water with other insects, the latter are soon found to be either mutilated or destroyed.

This Larva, on account of the transparency of the new skin, is in the best condition to be viewed by the microscope soon after it has cast its old skin, as the whole interior organization is then visible. This insect is furnished with two strong jaws, like a pair of bent pincers, which move horizontally and cross each other when closed. Their color is a bright chestnut, deepening in tint towards the points, which are hard and sharp. The animal seizes its prey with these instruments, and forcibly drawing it towards itself, makes an incision on the body and sucks out the juices. Unless its prey is of great strength the Larva does not kill it before eating, but, seizing on any part within its reach, devours it while its victim is alive. Having finished this portion, it turns the insect round and feeds upon a fresh part, and thus continues its repast until the whole of its prey is consumed, except the skin.

If the object of attack is a strong, aquatic animal covered with a shell, the Larva either seizes and holds it firmly grasped until it is exhausted by its efforts to escape; or, springing at it from time to time, cuts off its limbs in succession with its powerful nippers; then, turning the disabled animal upon its back, the ferocious creature pierces the mutilated body, and imbibes its contents. In respect to the other parts of the animal, feelers branch out from the head, each composed of four pieces, connected by joints. On either side of the head is a cluster of eyes, six in each cluster; in some species they are arranged in a circle at equal distances from one another, while in others three or four are united in a group with the rest a little separated from them. The organs of respiration are seen in their greatest development at the head, and their course from thence to the tail, through the entire length of the animal, may readily be followed. At the tail the different parts unite and terminate; they are not simple in their structure, but numerous lateral subdivisions are thrown out between the extremities of the main organs.

The entire body of the animal is composed of rings, decreasing in size from the head to the tail, which is forked, and consists of two strong spines with smaller ones branching from their sides. When one of these is destroyed, Pritchard observed it to be replaced by another, which seldom, however, attains the size of that which is lost. The Larva is provided with six legs, each consisting of three joints; bristling with formidable spines, fringed with hairs, and armed at the extremities with strong claws. The front part of the head, immediately below the jaws, is furnished with an arrangement like the teeth of a saw; but whether this apparatus is really composed of teeth has not yet been ascertained. This Larva feeds indiscriminately upon all kinds of aquatic insects which it can master, and is itself the prey of the larger water-beetles. When kept together by themselves, without their appropriate food, they attack and devour one another. If they are confined in separate vessels for a few days and then put together, they soon engage in fierce conflicts, seizing each other with their formidable jaws whenever a favorable opportunity occurs, and displaying the greatest courage; the assailant sometimes engaging another of twice its own size. These creatures move very swiftly through the water, occasionally rising to the surface for the purpose of breathing. They sometimes hold their tails above the water to attain the same end, admitting, while in this attitude, fresh portions of air into the respiratory organs by means of the orifices near the lower extremity of the body. As these animals gradually increase in size they become more and more inactive, and are often infested with the Bell-shaped animalcules. If this is the case the animalcules become exceedingly numerous, and when the Larva arrives at maturity, the surface of its body then appears to the naked eye to be covered with a fine down, which is nothing else than a vast collection of Bell-shaped animalcules.

The Lurco, or Glutton.—This aquatic animal resembles a caterpillar in form, and being transparent to a certain degree, affords an excellent object for the microscope, the internal structure of the creature being clearly discerned under a good light. The Lurco is usually found in collections of water where grass and weeds are lying partially decayed. When the day is bright they rove upon the surface of the water, but cluster together at the bottom in cloudy weather. No difficulty is experienced in preserving them alive for months in vessels of water, where they increase rapidly both in size and numbers, if bountifully supplied with their accustomed food. Pritchard states, that having caught in the month of June a number of specimens, which were nearly *one-fifth* of an inch in length, he kept them in a vessel holding about three quarts until the month of October; they had been plentifully provided with monoculi, and by this time had become very numerous, congregating together in masses of considerable size, and some of the larger individuals had grown to the length of *three-fifths of an inch.* The Lurco is not possessed of feet, but is furnished with small tufts of hair set along its sides. Its mouth is fringed with hairs, and when open has the shape of a pear. The gullet, connecting the mouth with the first stomach, is capable of instantaneously expanding to a great extent: it is never completely closed, and the prey of the Lurco, which it always swallows alive, may frequently be discerned moving about in the first stomach and seeking to escape through the mouth. The whole body of the creature is divided into a number of stomachs, separated from each other by a transparent ring of muscles, which expands and contracts to a considerable extent.

The Lurco is endowed with very strong digestive powers, for its favorite food is monoculi, which are covered with a hard, shelly case. These it swallows entire with great voracity, filling itself to repletion, when it remains torpid like the boa constrictor, as the process of digestion is advancing. A Lurco of an average size has been known to devour *seven monoculi* in the course of half an hour. At the end of this time five of them were seen moving in the first stomach, and the remaining two were lying in the second nearly dead. This voracious creature often swallows monoculi whose diameter is often longer than the ordinary width of its own body. In the figure three of its victims are seen within the body of the animal.

Eels in Paste.—If a paste is made of flour and water, and kept for a few days, its surface will be covered with a collection of minute, light brown animals resembling eels in shape. Not the slightest indication of life can be found in the paste when freshly made; indeed the heat to which it has been subjected in boiling would inevitably destroy any previous vitality; and yet but a short time elapses ere it swarms with countless living forms, whose existence, under ordinary circumstances, terminates only with the entire consumption of the paste upon which they feed. Adams, in his work on the Microscope, remarks, that there are *four* kinds of eels found in paste; that during the fall and winter they are *oviparous*,* and the young eels may then be perceived proceeding from the eggs; but at other times they are *viviparous*. The chief figure in drawing 112 is a magnified view of a full grown eel of the first kind. The position of the mouth is denoted by the letter *a*, and so perfectly distinct is this organ, that under the microscope it can be seen in motion as the animal feeds upon the paste; *c*, *c*, and *c*, are light brown particles of matter, which are found in the interior of the animal; *d*, *d*, *d*, &c., are young eels in the same situation.

When a full grown eel is cut in two, the young eels and the brown particles are at once expelled. The result of such a dissection is shown in the group around the central eel, where the individuals are magnified to the same extent as the parent animal. The smaller eels, found on the surface of the paste, and which have not arrived at maturity, exhibit the same appearance as the specimens composing this group, and display no internal organization, which only becomes apparent as they advance in size. So prolific is this animal that it often contains, when mature, a hundred living eels at one time.

The second species, which is oviparous, is delineated in figure 113, and is both

Fig. 113.

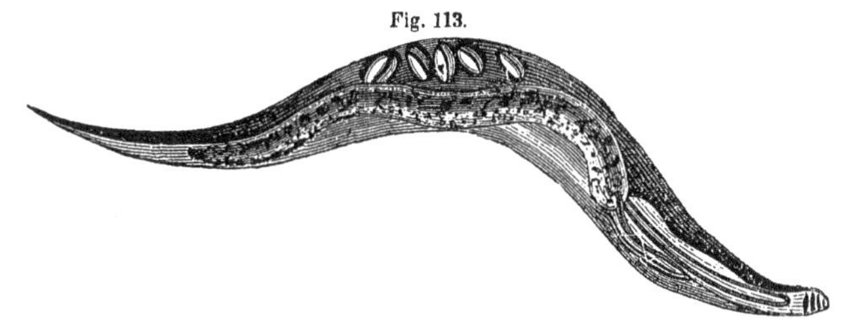

* *Oviparous*, producing young from eggs.

different in form and smaller than the first. The third sort is also oviparous, and is exhibited in figure 114. In figure 115 three extremely small eels are shown, which are specimens of the fourth class.

Fig. 115.

Fig. 114.

The manner in which these animals originate has given rise to much discussion, inasmuch as they are not only found to be viviparous; but are also said to be produced when the paste is covered. An opinion has therefore been entertained that their origin is spontaneous; a circumstance which would be at variance with all our experience. Dr. Ehrenberg, in respect to Infusoria, has experimented for many years with pure spring water, distilled water, and rain water; with and without vegetables, boiled and unboiled, and always with the following results: that when *open vessels were exposed to the air animalcules were discovered after a longer or shorter time*, according to the temperature, and other attendant circumstances; but if the vessels were *closed*, *infusorial life was rarely detected.* In view of these facts it may be inferred, that when eels are said to have been found in paste that was covered, that sufficient precautions had not been adopted to exclude it entirely from all access of the atmosphere, and that invisible eggs or germs, like those of Infusoria, were either contained in the air that floated at first above the paste in the jar, or were borne upon slight currents into the vessel, through minute apertures that escaped observation.

The fact that the eels of paste are *viviparous* cannot fairly be urged as an argument against their generation from a known cause, inasmuch as the same individual is both oviparous and viviparous; six young eels and twenty-two eggs having been found in a single specimen at one time. In their modes of increase they therefore resemble numerous species of Infusoria, which are not confined to one method of production, but multiply in various ways.

THE VINEGAR EEL.—If a small quantity of good vinegar is viewed in a wine-glass, by the naked eye, under a strong light, the fluid will generally be seen filled with slender threadlike bodies in rapid motion. These are the *eels of vinegar*, which, when studied under the microscope, are found to be larger than the Paste eel but not so thick; the tail is also smaller, more tapering, and moves with greater rapidity. Like the Paste eel they increase by eggs, and also bring forth their young alive. The eels of vinegar are finely displayed when a little reservoir is made between two narrow slips of glass, and the cavity filled with a few drops of vinegar. If the fluid is then magnified by the solar microscope, and its image received upon a large screen, hundreds of eels, apparently more than a foot in length, will be seen upon the screen in the highest state of activity, crossing and recrossing its surface, and darting and twisting in every direction. Sometimes if a small piece of mother floats in the vinegar, several eels will become entangled in it at the same time, and their rapid evolutions as they struggle to escape from this impediment affords an interesting spectacle. Their motions are evidently quickened by the glare of the sunlight, that falls upon them from the lenses, and which they endeavor to shun.

CHAPTER IV.

OF THE STRUCTURE OF WOOD AND HERBS.

I read His awful name emblazoned high
With golden letters on the illumined sky;
Nor less the mystic characters I see
Wrought in each flower, inscribed on every tree.
BARBAULD.

NATURALISTS have discovered by the aid of the microscope that all plants consist of two kinds of organic matter, essentially distinct, the *woody* portion and the *pithy* portion; and that the several parts of a plant, however differing from each other in form, texture, and appearance, are still composed of the same two substances, but varying in the proportion and arrangement. The woody portion has also received the name of the *vascular system*, while to the other division has been assigned the appellation of the *cellular tissue;* and these will now be described, so far as is necessary for the purpose of this work, without any design of entering fully into the subject of vegetable anatomy.

WOODY PORTION.—The woody part of a plant, whether herb or tree, is not solid, but is composed of a vast number of small tubes, extending from the roots, and ramifying through the stem and branches to every part of the plant; even the oldest and most compact species of wood is nothing else than a collection of vessels and cells, the sides of which consist of extremely thin and delicate membranes.

In the more highly organized animals, the vital fluid is distributed through appropriate channels by the action of the heart, throughout every part of the body. Near the heart these conduits are large, and few in number, but decrease in size and become less numerous as they are more remotely situated. In plants no such central fountain exists, but the fluids necessary for their life and development, entering from the soil through countless mouths at the roots, flow upward along the minute tubes of the plants, and are disseminated to every part where their presence is needed. The form of these tubes is generally cylindrical, and much difference exists in respect to their size. On account of the great minuteness of these pores it is extremely difficult to estimate their number correctly. An approximation to the truth may, however, be attained by first driving off the fluid that fills the pores, without destroying their figure, as is done in the preparation of charcoal, and then examining a cross section with a microscope. This method was pursued by Hooke, who numbered in a line, the eighteenth of an inch in length, no less than *one hundred and fifty tubes.* In an inch long there would consequently be (18×150) twenty-seven hundred tubes, and in a square inch (2700 × 2700) *seven millions two hundred and ninety thousand tubes.* The examinations of decayed wood, where the tubes were empty, led to the same result; and further corroboration was obtained by Dr. Hooke from the inspection of petrified wood, where the situation of the pores was very conspicuous. In woods that are remarkable for their compactness and density, the vessels or tubes are still smaller and more numerous within a given space. The largest tube observed by Hedwig, in the stem of the gourd, possessed an apparent size of one-twelfth of an inch, when seen through a microscope that magnified two hundred and ninety times. Its real diameter was therefore one-twelfth of an inch, diminished two hundred and ninety times: or the three thousand four hundred and eightieth part of an inch. If, therefore, within the extent of a square inch a collection of tubes like these occupied but one half of the space, no less a number than *six millions fifty-five thousand two hundred* would be comprised within this small compass.

Arrangement.—These tubes are not arranged singly throughout the trunks and branches, but are collected into small bundles; in the stems of herbs and of roots, each small bundle is composed of from thirty to one hundred tubes, and sometimes of many hundreds. In *herbs* the bundles are often placed at considerable distances from each other without any symmetrical arrangement, while in *trees* they are regularly disposed in concentric circles; and, when cross sections of wood are viewed through a microscope, are seen distinctly arranged in a great variety of the most beautiful and elegant figures. It was supposed by the earlier writers on vegetable anatomy, that the tubes which have just been described were of *two kinds;* the office of the first class being to convey *sap,* and that of the second to carry *air* to the different parts of the plant. The tubes were thence denominated *sap-vessels* and *air-vessels;* the latter class also received the name of *spiral vessels,* from the peculiar manner in which the tube is formed. It is now however believed by distinguished naturalists, that there are no vessels *exclusively* employed for the conveyance of air, but that all the tubes found in the woody parts of plants are sap-vessels of *one kind* only, and that in different plants, and in different circumstances and conditions of vegetation, these vessels or tubes are capable of assuming various forms, sizes, and characteristics; a circumstance which has led many observers to the belief, that they constituted several distinct species, which subserved different purposes conducive to the life and growth of the plant.

In figures 117, 118, 119, 120, 121, and 122, is delineated the sap-tube, under the several forms which at times it assumes. In figure 117, it appears to be covered with rows of small projecting knots. Stripes cross it in figure 118; while in figures 119 and 120, the spiral structure is distinctly seen, the tube being formed in the same way as a paper-lighter is made, by rolling a long and narrow surface spirally upon itself. In figure 121, the vessel appears to consist of a

Fig 117. Fig. 118. Fig. 122.

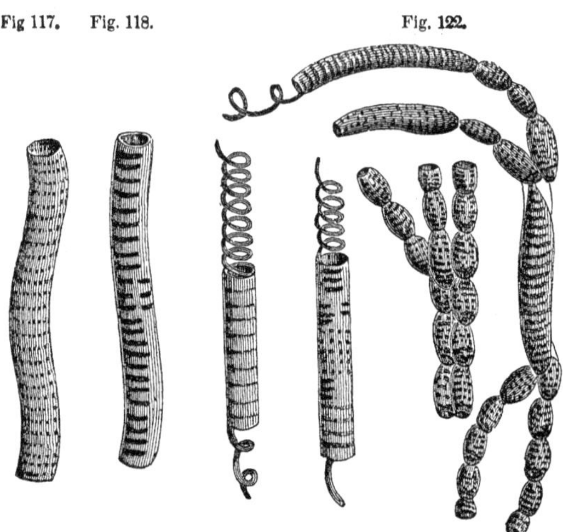

Fig. 119. Fig. 120. Fig. 121.

series of cups or beads strung together, the surface of the several parts being studded with minute projections. The spiral structure is again seen in the 122d figure, and the characteristics of figures 117, 118, and 121, are here likewise recognised.

CELLULAR TISSUE.—The tubes just described are bound together by a tissue filled with minute cells, which has thence been denominated the *cellular tissue.* It is a constituent part of every organ of the more perfect plants, and in many herbs forms the principal portion of their substance; while the lower order of vegetables, as mosses, mushrooms, &c., are said to consist of it entirely.

The appearance which this tissue presents is extremely diversified. At one time it is seen to be of a *loose, porous texture*, every part of which is transparent and succulent. Under other circumstances, it meets the eye in the condition of a *solid body*, the cells being so closely pressed together that the peculiarity of its structure is almost lost. In a third case the cells likewise vanish from another cause, for the tissue then spreads out into a membrane so extremely delicate and thin that all traces of their existence disappear. The cells or cavities of the cellular tissue are generally arranged in a direction opposite to that of the tubes of the vascular system, and are therefore displayed in the *longitudinal* and not in the cross section of a plant.

The forms of the cells are exceedingly various. In some plants they are of a globular shape, in others angular, but differing as to the number of sides; several kinds being triangular, others square, but the greater proportion exhibit hexagonal or six-sided figures.

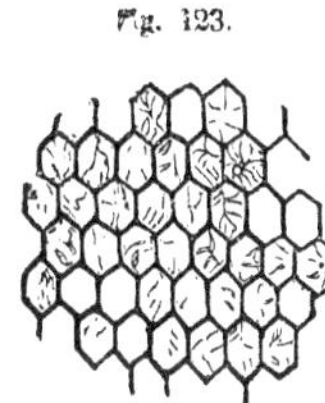
Fig. 123.

In figure 123, is shown, drawn from nature, a transverse slice of the cellular tissue of a sugar-cane, so thin as to display only one layer of cells, but a thicker slice of the same plant exhibits a second set of cells behind the first.

These cells vary greatly in respect to magnitude, and are represented by one naturalist as possessing twenty different sizes, ranging from those as large as a pea to others which are so minute as to require the aid of powerful microscopes to perceive them distinctly.

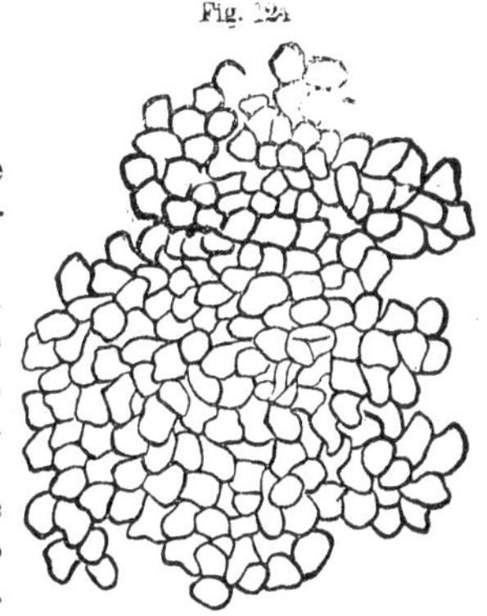
Fig. 124.

Hooke examined a thin section of cork, and found that no less than sixty cells were placed endwise in a line the eighteenth part of an inch in length; more than a million would therefore be comprised within the surface of a square inch.

In most plants the pores of the cellular tissue are readily discerned, but in cork they require to be highly magnified in order to be clearly seen. A thin slice of cork thus magnified is delineated in figure 124. The substance is seen to consist of an assemblage of minute cells formed of extremely thin membrane. The average size of the cell is about the *eight hundred and thirtieth part of an inch* in diameter.

PITH.—When a cross section of a tree or plant is viewed by the naked eye, it is seen to consist of *three parts*, the *pith*, the *woody texture*, and the *bark*. The size of the pith varies in different trees, in some being from two to three inches in diameter, and in others from five to six; and of all plants, herbs and shrubs have the greatest quantity of pith in proportion to the other parts. The pith is found to consist entirely of cellular tissue, and the cells are of various sizes. Those of the thistle appear under the microscope as large as the cells of a honey-comb; those of plum, wormwood, and sumach, are smaller, and the cells in the pith of the apple and pear are still less; while those of the oak are so minute that one hundred only equal in size a single cell of the pith of the thistle. The size of the cells is not proportioned to that of the pith, for in the plum, the pith of which is less than that of the pear, the cells are from four to five times as large; and the cells of the pith of the hazel, which is three times smaller than that of the holly, are ten times greater than those in the holly.

WOOD AND WOODY TEXTURE.—The second portion of the plant is the wood or woody texture; it encircles the pith, and consists, as already stated, of two parts; *bundles of tubes*, bound together by *cellular tissue*. In most trees the vessels are numerous, and when beheld in a cross section are seen to be disposed around the pith in concentric layers, and rays of cellular tissue to run from the pith to the bark, diverging like the spokes of a wheel from the axle. This arrangement is seen in drawing 134, which represents a cross section of part of an ash branch three years old. The numerous vessels of the wood, which are denoted by small circles, are here seen occupying the space D C I K; from the pith I K L to the bark A B C D; and the insertions of the cellular tissue are indicated by the lines that run from the pith outward like the sticks of a fan. These insertions of tissue pass through the substance of the wood, and are much diversified in size in different woods. In pine they are of a medium size, and in pear and holly extremely small, but no uniformity in this respect is observed in the same wood, for in holly, hazel, pear, plum, and oak, they are very unequal; some in the holly being four or five times thicker than the rest; while in the plum many are six or seven times greater than others, and in the oak ten times at the least. In trees like the palm, the vessels of the vascular system are by no means so numerous as in other woods, and being necessarily placed at a greater distance from each other, do not present that symmetrical radiated appearance which sections of common trees exhibit; but the bundles of tubes are promiscuously scattered amid the cellular tissue. This is evident from a glance at figure 125, which represents a cross section of the palm, the dark spots indicating the position of the vessels, and the lighter the cellular tissue. In the case of herbs, to a great extent, the cellular tissue forms the chief portion of the plant, and the vessels of the vascular system are but few in number. When a cross section is viewed they are seen in bundles dispersed through the cellular tissue, at considerable distances from each other; they are, however, symmetrically arranged, and in the same species of plant always maintain the same position; the vessels being situated at the same relative distance from each other and from the centre of the pith.

Fig. 125

BARK.—This envelope, which encircles the wood, is composed of *two* parts, the *true bark* and the *outer* skin which covers it; both of which are made up of vessels and cellular tissue like the wood. The tubes or vessels belonging to the bark are denominated *proper vessels*, and are filled with the fluids peculiar to this portion of the plant. In some herbs these vessels often cluster together in separate columns, and are arranged in a circular form; in others they present a radiated appearance. In trees, the tubes are more distinct, and assume a greater regularity in their disposition.

They are usually found near the inner margin of the bark, next to the wood, and when viewed in the direction of their length, present an appearance like network. In the bark the vessels are found to possess different sizes as well as in the wood. In the pine, for instance, the tubes containing the turpentine exceed the common sap-vessels in magnitude by three or four hundred times, and are surrounded by a ring of smaller tubes. In drawing 143, the proper vessels of the bark containing the milky juice of the sumach are arranged in arched clusters, each cluster consisting of several hundred distinct tubes, and so small are these tubes that a single turpentine vessel of the pine sometimes vies in magnitude with an entire arched cluster of the sumach.

Fig. 126.

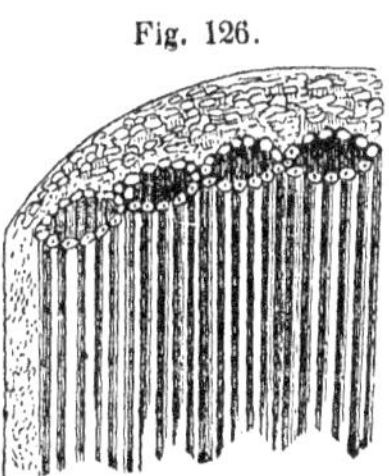

In figure 126 the vessels of a certain species of pine are displayed, where the large turpentine tubes of the bark are seen encircled by a ring of smaller tubes; the woody part of the tree having been cut away, so that the longitudinal as well as the transverse structure can be clearly seen. Dr. Hill discovered by close investigation, that each of these large tubes was nothing more than an opening in a bundle of small tubes, and he remarks "that if we conceive any small bundle of tubes to be opened in its centre, and the vessels driven every way outward until they are stopped by the substance of the bark, we shall have an idea of the structure of this larger vessel; which is nothing more than a great cylindrical tube, passing through the centre of a bundle of smaller ones." This structure is plainly perceived in the figure.

The vessels containing the fluids peculiar to the bark are often found dispersed through the wood from bark to the pith. Thus, in the fir and pine, the turpentine and gum-tubes are seen in the wood, arranged in a circle around the centre, in nearly the same way as the sap-vessels. These vessels are regarded by naturalists as having once belonged to the bark, which, changing into wood by the natural growth of the tree, in the manner soon to be explained, and becoming encased annually in successive layers of wood, was gradually removed farther and farther from the exterior surface of the tree. The skin or rind of the bark when taken from young shoots, appears in most cases to consist of a single layer, but in many kinds of wood it is found to be complex; as in the case of the white birch, in which it consists of distinct layers that are readily separated from each other, amounting not unfrequently to sixteen or eighteen in number.

In some trees the layers are still more numerous, for Ulloa speaks of a tree in Peru, from the rind of which he peeled off no less than *one hundred and fifty* envelopes; when, tired of his task, he refrained from counting the remainder, as the layers he had detached did not constitute more than half the thickness of the rind.

THE MODE OF GROWTH IN THE TRUNK AND BRANCHES OF TREES.—Commencing on the outside of the tree, the exterior covering is the skin or rind, consisting, as has already been stated, of several distinct layers. Beneath this is the bark, composed of cellular tissue and bundles of tubes or vessels running longitudinally; and at first parallel to each other. When a cross section is made of a shoot of a year old, only a single ring of vessels, or cluster of vessels arranged in a circle, is found surrounding the wood, and this, with the tissue in which they are enveloped, constitutes the bark of the plant. During the second year, a new layer of bark with its vessels and tissue grows within the former, and next to the wood; and every successive year a new layer of bark is thus added: the entire thickness of this envelope being constituted of a series of layers, increasing in number *from within.*

Next to the bark, the wood is found consisting likewise, as we have seen, of vessels and cellular tissue, and the cross section of a plant or shoot of one year's growth, exhibits the wood arranged around the pith, and composed of a ring of vessels banded together by cellular tissue. The growth of the succeeding year gives rise to a new ring of vessels outside of the first ring, and beyond this second ring, a third circle of vessels appears during the third year. The wood of the tree therefore increases from *within outwards*, in a direction contrary to the growth of the bark; and consists of a series of rings, equal in number to the years indicating the age of the tree. The outer ring is whiter and more full of sap than the rest, and has received from this circumstance the name of *sap-wood.* In the last annual layer of wood and bark, by which the trunk is increased in size, the sap-wood and new bark are in contact; but the layer of the next year pushes up between, and separates them by its entire thickness. Every year a new layer is thus interposed in the midst of the others, the last formed layers of wood and bark touching each other, while the oldest are the most widely separated; the first ring of wood directly enclosing the pith and the first envelope of bark constituting the outer surface.

The layers of wood annually formed are not simple in their structure, but are each composed of a great number of other layers. These delicate membranes can be distinctly perceived in the oak by the aid of a common magnifying glass, when the branch or shoot is cut obliquely. By macerating the rings of wood in water, Du Hamel was enabled to separate an annual layer into a vast number of primary layers, which were thinner than the finest paper; and he afterwards discovered by experiment that these primary layers, constituting any annual ring, were formed in *succession*, during the period of growth and vegetation in the year to which the ring belonged. So that however curious it may seem, it is still true, that not only does the relative thickness of each annual ring indicate the comparative fruitfulness of every year of the existence of the tree, but each of the primary layers composing the several rings tells, by its thickness, of the comparative vegetative energy of each week and day of the particular season to which it belongs; and thus every tree becomes a record of the fertility of that period of time during which it lived and flourished. The branch possesses exactly the same structure as the trunk.

PLATE II.

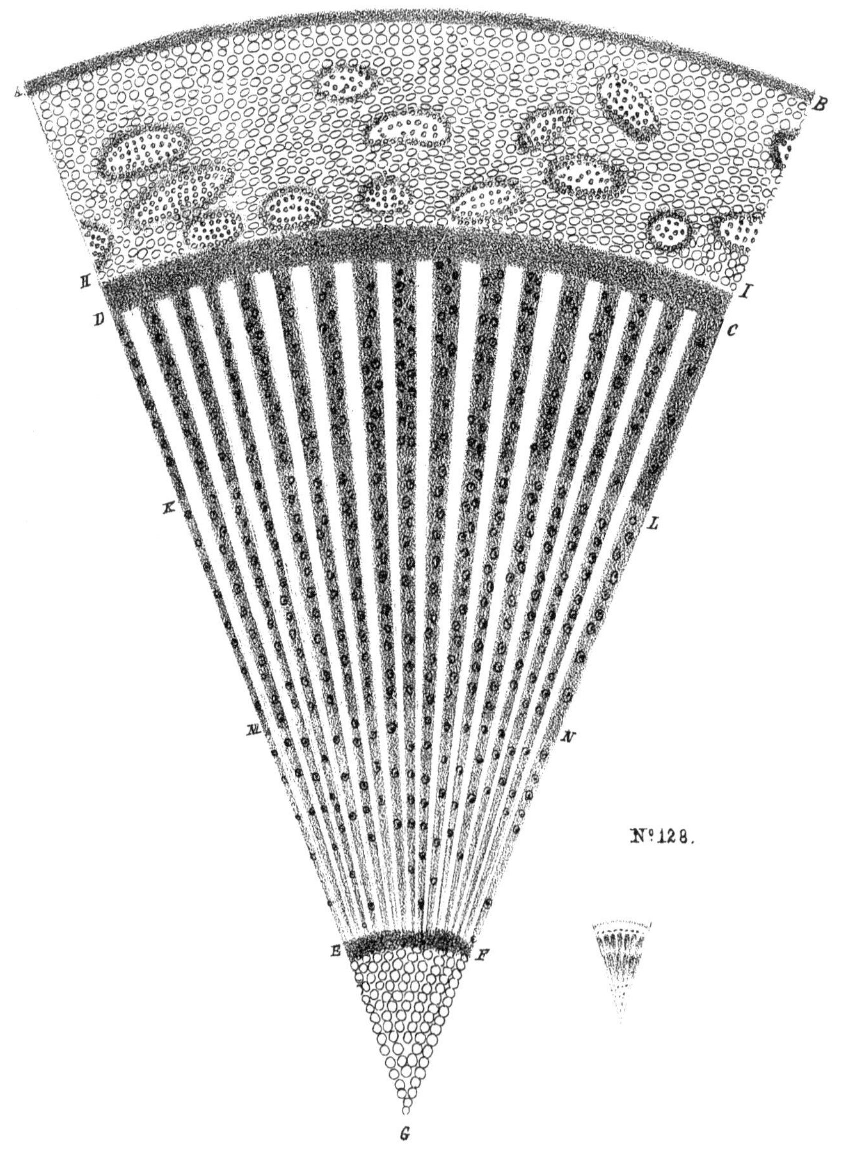

Nº128.

SECTIONS OF WOOD.

Pear Tree.—In drawing 128 is presented one-eighth part of a section of a branch of a Pear tree, both magnified and of its natural size. It exhibits, of course, the same general features as the section of the Holly, but varies in some particulars. The rows of vessels stretching out from the pith to the bark are less broken than in the Holly, and the rays of the cellular tissue are more regularly arranged; branching out at equal distances from each other, and presenting, with the numerous vessels dispersed throughout the wood, a remarkably elegant figure. The three rings of the true wood E F M N, M N K L, and K L D C, denoting the age of the branch, are distinctly marked; and beyond them the sap-wood occupies the space D C H I. The bark is comprised within the space H I A B, and is formed of the minute cells of the cellular tissue, interspersed with numerous figures of an oval shape. These last are clusters of proper vessels of the bark, and are rounder in form, and more numerous, the nearer they approach the wood.

The Hazel.—A section of a Hazel branch, when magnified, exhibits a figure of exquisite beauty and symmetry. In drawing 129, an eighth part of an entire cross section of a bough three years old is faithfully delineated. The bark, which is included within the space A B C D, is enriched with clusters of vessels of various shapes and figures. The skin of the bark is represented by the ring A B, to which succeeds a broad band Q Q, consisting of the cells of the cellular tissue; within this band is another ring H I, composed of sap-vessels. The space H I D C is filled partly with cellular tissue and partly with sap-vessels, in pear-shaped and semi-oval clusters, alternating with each other at equal intervals. The wood extends throughout the space D C E F, and is divided into regular and equal compartments by great radial insertions of cellular tissue; and these compartments are still further subdivided by more delicate and minute rays of tissue running from the pith to the bark. Throughout the wood spiral vessels are profusely scattered, and are found to be most numerous near the bark, or in the growth of the last year, D C K L. The growth of the second year is comprised within the space K L M N; while that of the third year is represented by the space M N E F. The size of the pith, E F G, is small compared with that of the wood; but its cells are much larger than those belonging to the pith of the Holly. The small figure at the bottom of the plate represents the large section of its natural size.

English Oak.—A section of Oak varies in some respects from the several kinds of woods which have already been described. The vessels of the bark are arranged in different ways. Through the middle of the bark extends an unbroken arched band of vessels, while a row of large oval clusters, standing at equal distances from each other, intervene between it and the wood. The vessels of this inner row are of a peculiar nature, and are termed by Grew *resiniferous* vessels, since they contain a thick juice of the Oak which is not sap, but bears the same relation to it in the Oak as the turpentine of the Pine to the sap of that tree. The rays of cellular tissue, diverging from the centre to the bark, are in this tree divided into *two kinds* as regards size. The first are broad insertions; which, for the most part, are of the same size, and are disposed around the centre at regular intervals; the second are the finer radial divisions, which in like manner are uniformly arranged and occupy the spaces between those of the first class.

In the section of the Pear, all the dividing lines are seen proceeding from the centre, but in the Oak and some other trees numerous white waving lines run across the radial divisions. These undulating rings constitute, in a great measure, the beauty of the Oak, and are considered by Grew as *sap-vessels*, which once existed in the bark, but in process of time became condensed and hardened into wood.

WHITE OAK.—A section of the common White Oak, magnified one hundred and thirty-seven times, is delineated in figure 131. The broad band *a a* is one of the insertions of cellular tissue radiating from the pith; it is exceedingly compact, for no pores can be detected within it when subjected to this high magnifying power. Narrower insertions of cellular tissue *b b*, &c., traverse the wood in the same direction in irregular waving lines. The spiral vessels *d c*, &c., are scattered in considerable numbers throughout the wood, occupying a large proportion of its space. They vary much in size, the smallest, as *d* for instance, not being more than one *four-hundredth of an inch* in diameter; while one of the largest, as *c*, measures not less than the eighteenth part of an inch across it.

Two sets of these large spiral vessels are seen in the figure, which, like those beheld in the section of the English Oak, are arranged along the inner margin of each annual ring of wood; the distance between the two clusters being the thickness of one year's growth of wood. The wood itself appears like lace-work, being filled with minute pores, varying in diameter from *one twelve hundred and fiftieth part of an inch* to *one twenty-five hundredth.*

PLATE III

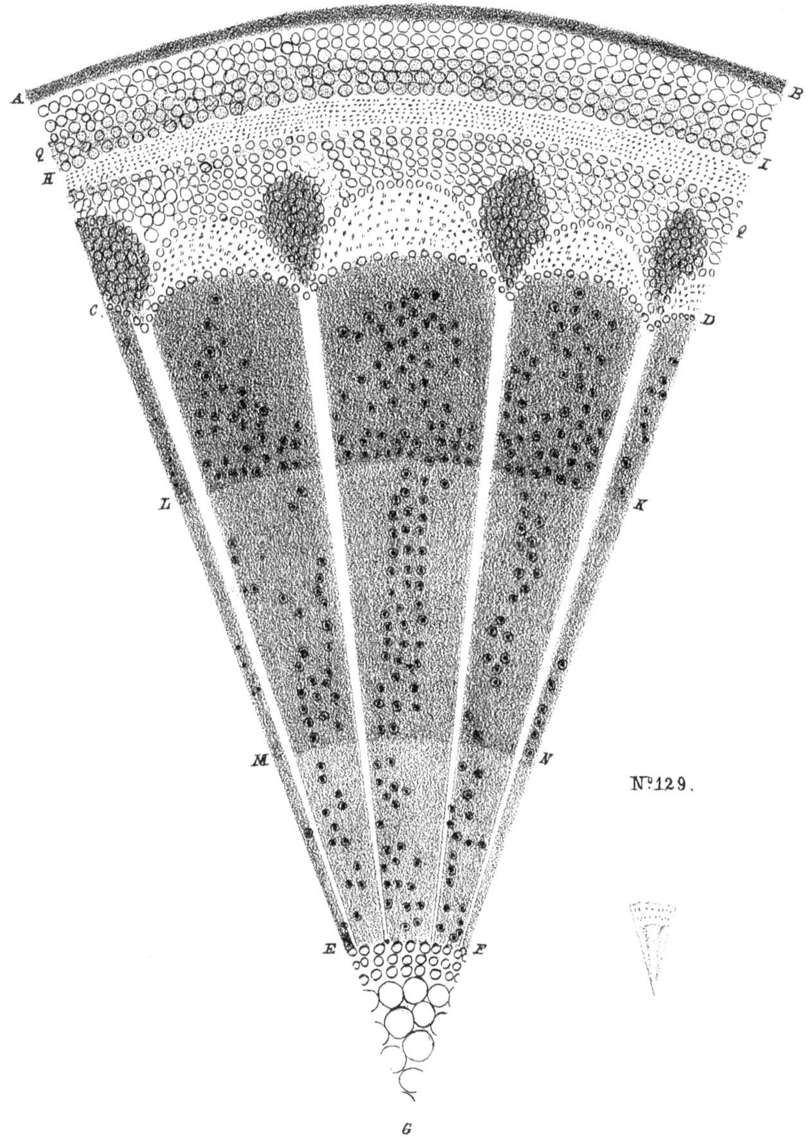

Fig. 131.

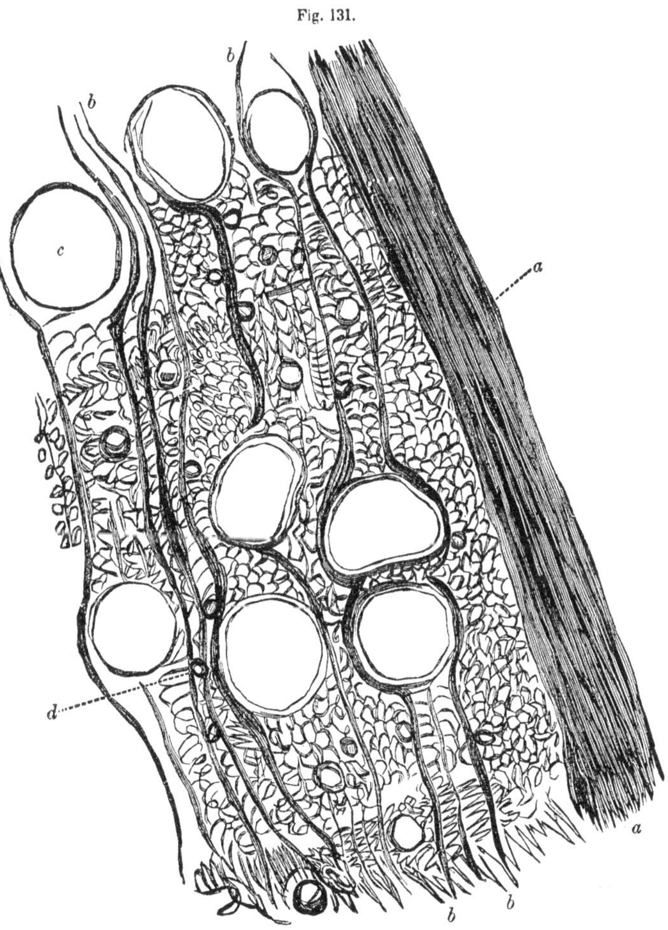

ELM.—A section of an Elm branch which appears, is an exceeding rich figure, and the several divisions are boldly defined. The skin of the bark possesses considerable thickness, and the pores of the cellular tissue belonging to this integument are exceedingly small.

Throughout the bark, bundles of the proper vessels are seen profusely scattered in oval or egg-shaped clusters, and beyond it the sap-wood. In the true wood, the annual rings are very distinctly marked. The vessels dispersed through the wood differ very much in size; the larger, disposed in circular bands, are arranged on the inner margin of every annual ring, while others are scattered promiscuously throughout the wood, and are more numerous near the centre of the section than in the more recently formed wood towards the margin. The more minute vessels are seen stretching in delicate and broken chains across the rays of cellular tissue, emanating from the centre; the various positions of the rays being indicated by the white lines in the figure running from the pith and penetrating for some distance into the bark. The rays of cellular tissue possess great uniformity in their respective thicknesses, as well as in the intervals by which they are separated from one another. The rays are usually arranged at equal distances from each other.

ENGLISH WALNUT.—In drawing 133 is displayed a magnified section of a branch of English Walnut, four years old, and which presents a most beautiful configuration. A B indicates the position of the skin of the bark, and the latter envelope, with its cellular tissue, and proper vessels, is comprised within the space A B C D. The proper vessels collected together in round clusters are distributed in two circular rows H I and R S, deeply situated within the bark. The wood is included in the space D C E F. The first annual ring extending from E to O, the second from O to M, the third from M to K, and the fourth, including the ring of sap-wood, P D, from K to D. The sap-vessels distributed through the wood are not numerous, but their size is comparatively great, and, as in the Elm, they are grouped more thickly together near the pith, the cells of which are quite large compared with those of the pith of the Elm. The pith itself is also much larger than in many other woods of the same age. The radial lines of cellular tissue in the Walnut observe no uniformity in respect to their relative thickness, as is the case of the Elm, neither are they arranged at equal distances from each other.

But the most remarkable peculiarity in the Walnut is the broad *white arched bands* running across the rays of cellular tissue, four of which are exhibited in the figure before us; in the Elm they are also seen disposed in a similar manner but much narrower. Their existence is attributed to the same circumstances that cause a similar appearance in the Oak; namely, the greater compression of the cellular tissue where these bands occur than in other portions of the wood.

ASH BRANCH.—A cross section of part of an Ash is a handsome object when magnified to a considerable extent. The bark consists of an infinite number of cells formed by the cellular tissue. Within the bark, next to the skin and nearest to the wood, clusters of minute vessels extend in two circular rows from side to side. The arrangement of the radial insertions of cellular tissue is very beautiful; the rays diverging from the pith to the bark at equal distances from each other, and maintaining, nearly always, the same size. The position of the large spiral vessels

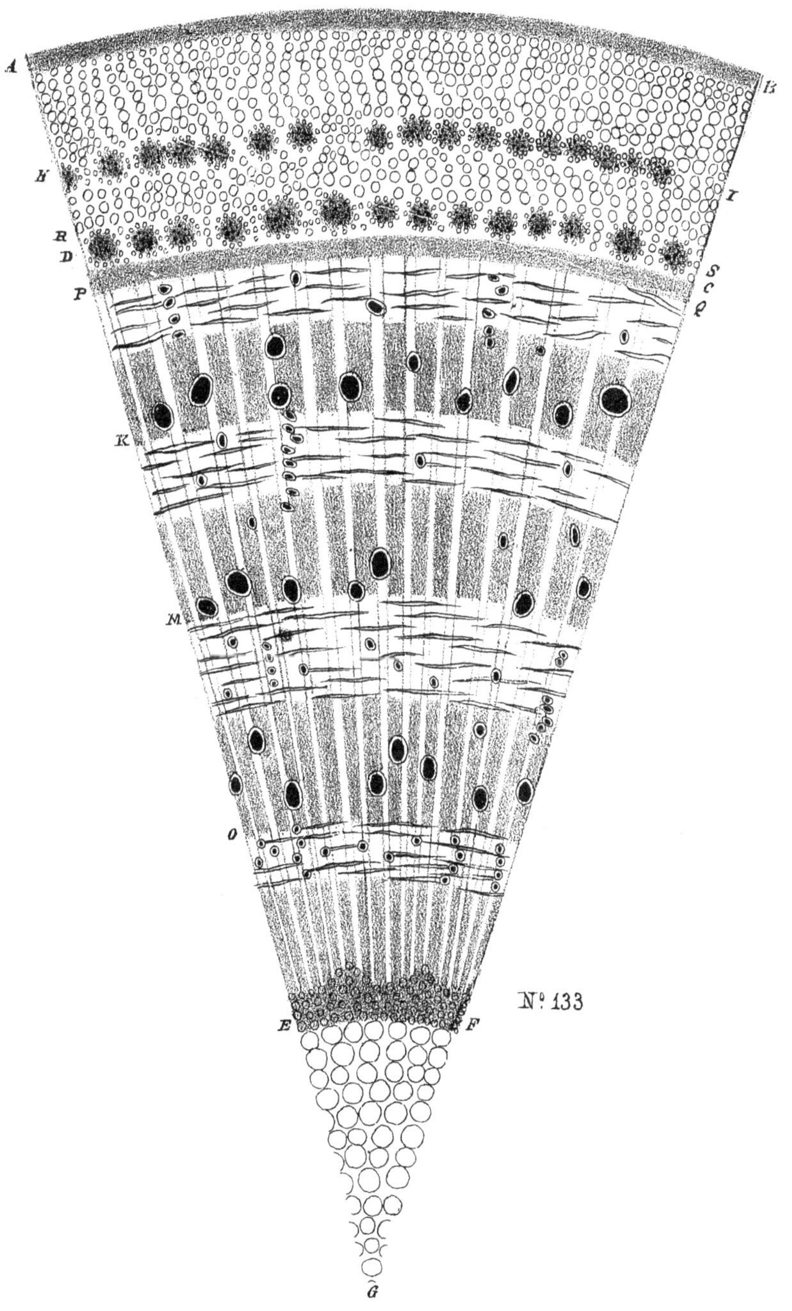

Nº 133

in the wood is very distinctly marked, gathering in arched bands near two divisions of the annual growth.

MAPLE.—A cross section of the firm wood of the Maple is presented in figure 135, highly magnified. The strong dark lines are the rays of cellular tissue, emanating from the centre of the trunk of which this section represents a part. The large oval openings are sections of the spiral vessels which run lengthwise through the trunk, and the rest of the figure shows the true wood filled with minute pores, whose size does not exceed the actual measurement of *one twelve-hundredth of an inch* in diameter.

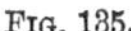
FIG. 135.

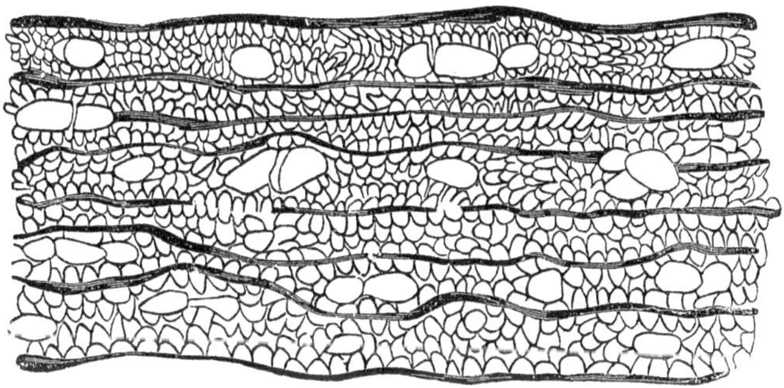

DOGWOOD.—A magnified cross section of Dogwood is delineated in figure 136. This wood is very hard and firm in its texture, and the smaller pores are much more minute than those of the maple and other lighter woods. In the specimen exhibited, a multitude of fine oval pores are seen scattered throughout the wood, the largest of which does not exceed one *two-thousandth of an inch* in diameter, and the smallest is not more than one *three-thousandth of an inch.*

The large openings *y y*, &c., are spiral vessels, having a diameter of about one *two hundred and fiftieth part of an inch.* The heavy and boldly defined lines running lengthwise of the figure are the rays of cellular tissue which proceed from the pith to the bark.

Fig. 136.

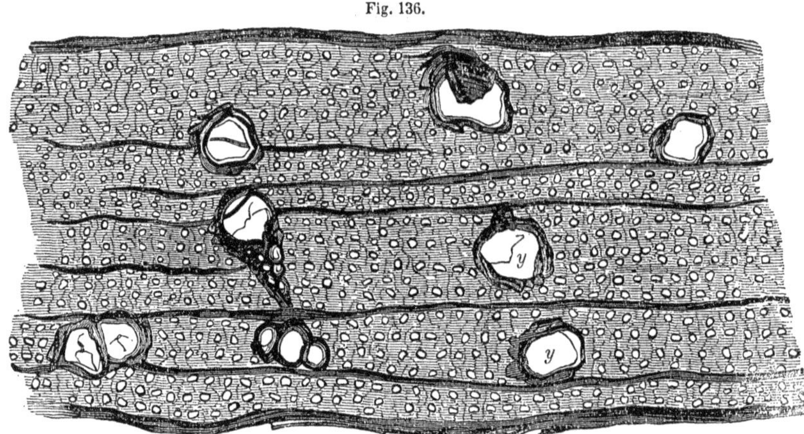

WHITE PINE.—A transverse section of White Pine, magnified four hundred times, is presented in figure 137. The cause of the lightness of the Pine is seen

a Fig. 137. *a*

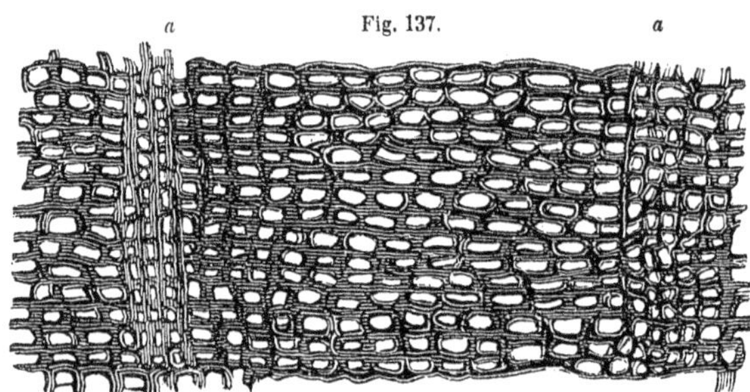

at a glance; for the wood is as full of openings as a piece of lace-work, and consists of nothing but a web woven of the fine fibres of the cellular tissue. Across the figure at the points *a a*, bands of cellular tissue are beheld, stretching from side to side, and the structure is here more compact than in any other part ot the wood. These divisions are portions of concentric annual layers formed by the compression of the cellular tissue.

Fig. 138.

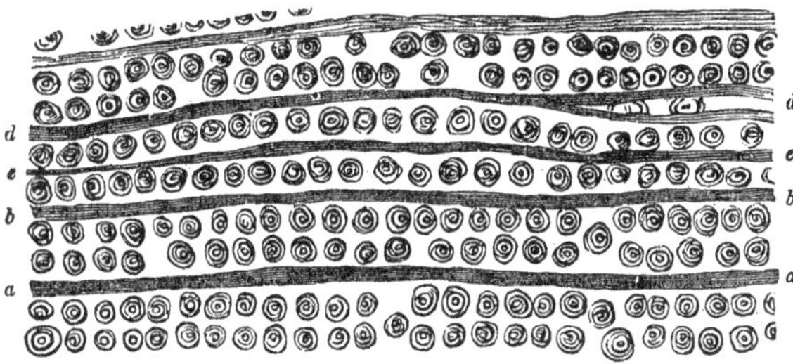

A longitudinal section of the same tree is delineated in figure 138, as it appears when magnified likewise four hundred times. The structure is exquisitely beautiful: the straight lines *a a*, *b b*, *c c*, *d d*, &c., are sections of the *sides* of the vessels or tubes which run lengthwise of the trunk, the ends of which are disclosed in the woven lines of the last figure. The sides of the vessels are seen studded with rows of small circular disks which have received the name of *glands;* each disk having a small circular ring around its central point. The form of the disk is not always the same; it is generally circular but frequently oval; and when closely arranged together they assume an angular shape. In some species of Pines the disks run through the vessels in single rows; but in others, as in the case of the White Pine, they occur, as is obvious, both in single and double rows. It is a remarkable fact, that throughout the entire genus of the living, true Pines, no more than two rows of disks are ever found in the longitudinal section of a single vessel, and that when a double row occurs, the corresponding disks of each row are placed side by side. The vessels are sometimes found without disks.

A class of cone-bearing trees, allied to the Pines, is known by the name of Araucaria. It includes some of the loftiest living trees, and the well-known species that grows in Norfolk Island, near New South Wales; and which bears the name of the Norfolk Island Pine. This class possesses certain peculiarities of structure which are at once detected by the microscope, and distinguish it from the true Pines.

In figure 139 a longitudinal section of the Norfolk Island Pine is displayed, magnified to the same extent as the two preceding figures. The disks that cover the sides of the vessels are here arranged in double and triple rows; and in the Araucarias the rows belonging to the section of a single vessel vary in number from one to four.

Fig. 139.

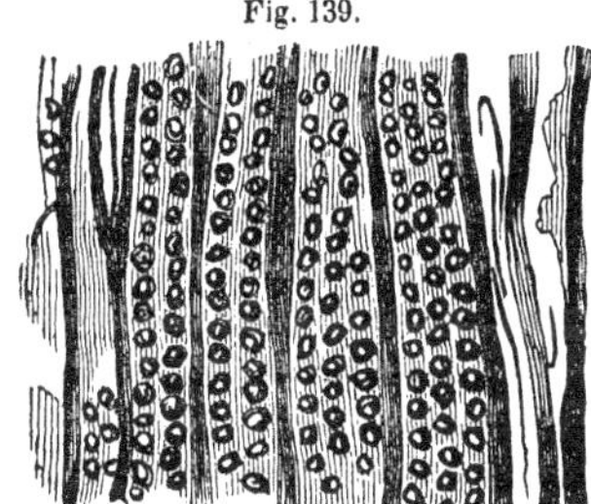

Another peculiarity is also perceived in the shape of the disks, which, instead of being generally circular like those of the pine, are for the most part bounded by straight lines. The disks also of both the rows in a double row, are not placed side by side, but always alternate with each other. The first disk of one row, for example, being placed opposite the vacant space separating the first and second disks of the adjoining row. The disks of the Araucarias are not more than one half the size of those belonging to the real pines, and are so numerous that Mr. Nicol counted as many as fifty in a row, one-twentieth of an inch in length. The diameter of a disk, allowing that they touched each other, could therefore not exceed the *thousandth part of an inch;* a magnitude of vast extent compared with the thickness of those infinitely slender fibres, which, woven together into an exquisitely delicate tissue, form the partitions of the numerous cells of the trunk.

Fig. 140.

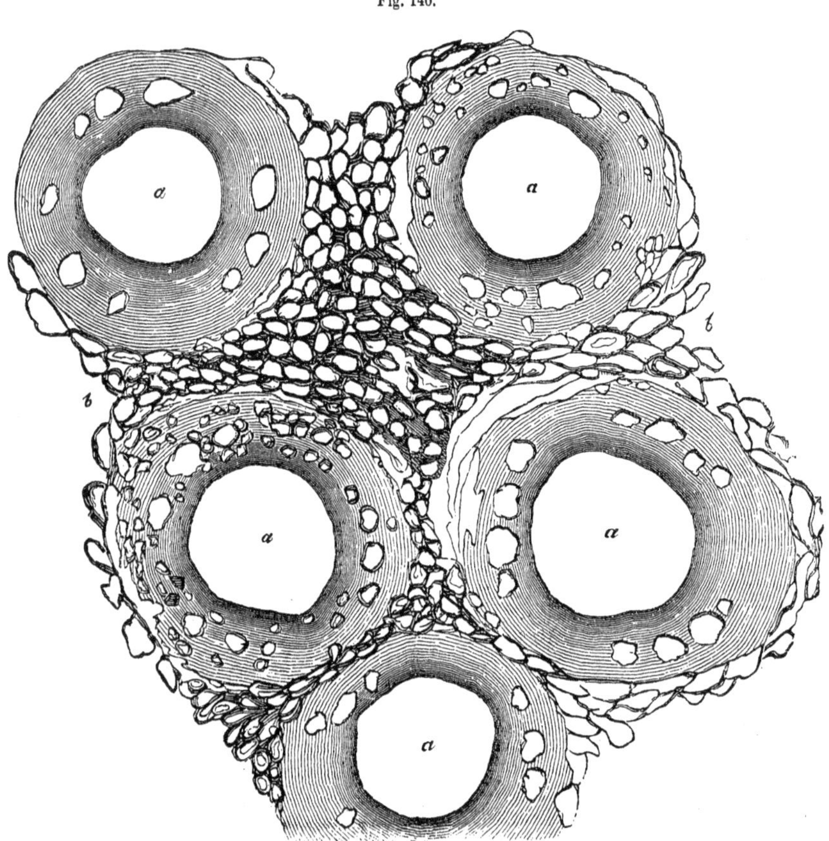

MALLACA.—A portion of a cross section of a species of Bamboo, found in the Mallaca isles, from whence its name is derived, is exhibited in figure 140, magnified one hundred and fifty times. The five large rings are bundles of vessels running lengthwise through the trunk, the vessels of each forming by their mutual arrangement five tubes (*a a a a a*) of considerable size, which traverse the bundles through the centre from end to end. So porous is the stalk of the Mallaca, that by applying the mouth to one end of a stem several feet in length, a lighted lamp placed at the other extremity can almost be extinguished by the impulse of the breath, blown through the stalk; nor is this surprising, for the large vessels (*a a a a a*) are *one-seventieth part of an inch in diameter.* The vessels composing the rings are evidently far inferior in size to the tubes; and upon actual measurement the greatest are found not to exceed *one-five-hundredth of an inch in diameter.* An elaborate net-work of cellular tissue (*b b*) twines around and among the cylindrical clusters of vessels, and binds them firmly together with its delicate but strong cordage.

In figure 141 is exhibited a drawing of a cross section of the common Larch,

Fig. 141.

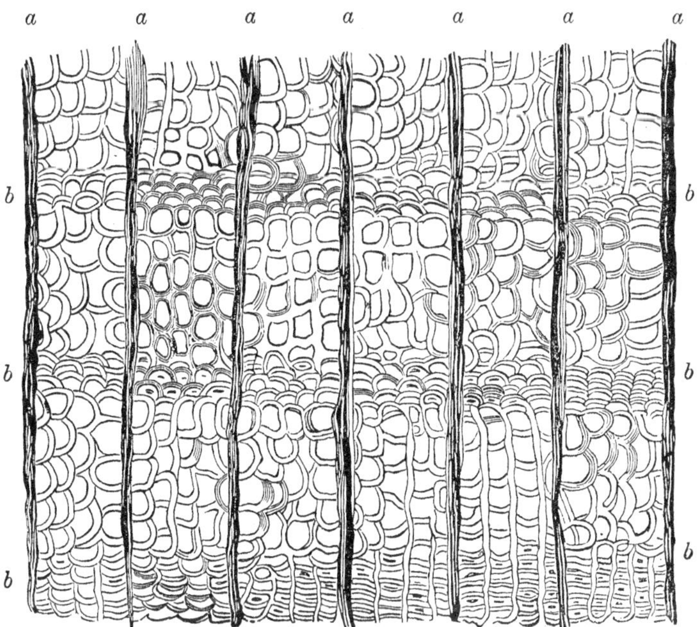

greatly magnified. The strong lines *a a a a a a*, are the rays of cellular tissue which proceed from the centre to the bark, but on account of the smallness of the sections magnified, their divergence from each other is too small to be detected. The remaining portions of the wood consist of a perfect chainwork of cells, formed of the cellular tissue. These cells are compressed, like those of the Oak, at regular intervals, as shown at *b b b b b b*, in directions apparently at right angles to the radial insertions. The width of the cells, where they are not compressed, is about *the twelve hundred and fiftieth part of an inch*, and the thickness of one of the radial insertions, as *a b b b*, is a little less than the *nineteen-hundredth part of an inch.*

WHITEWOOD.—The elegant structure of a transverse section of Whitewood, as revealed by the microscope, is displayed in figure 142. The four heavy lines,

Fig. 142.

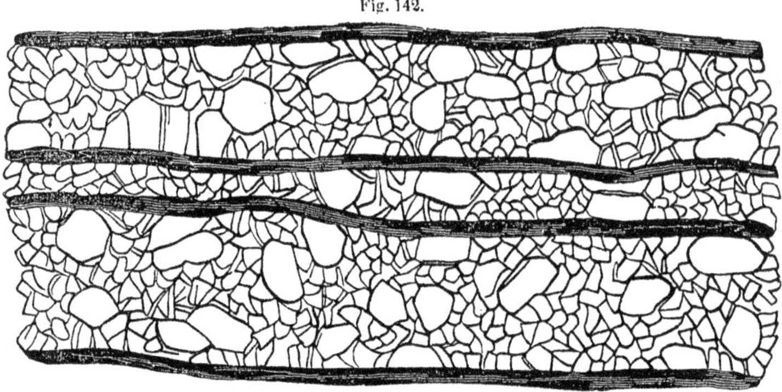

running in a parallel direction with each other, are some of the radial insertions of cellular tissue, and in thickness do not exceed *one-twelve hundredth part of an inch.*

The large openings scattered throughout the wood are sections of the spiral vessels, the diameters of which, when largest, measure only the *three hundredth part of an inch.* The rest of the space is filled with a most exquisite network of fibres, the meshes of which are angular in form, the whole surpassing in the delicacy of its texture, the fabric of the finest laces.

SUMACH.—A singular figure is exhibited in drawing 143, which represents one eighth part of a cross section of the common Sumach of one year's growth. The bark occupies the space A B K L, the pith that of E F G, while the wood, including the sap-wood K L C D, is comprised within the limits K L E F. The bark is covered with a down of fine hairs, which when magnified, fringe the section with the thorn-like figures *a a*, as shown in the drawing. Most of

PLATE V.

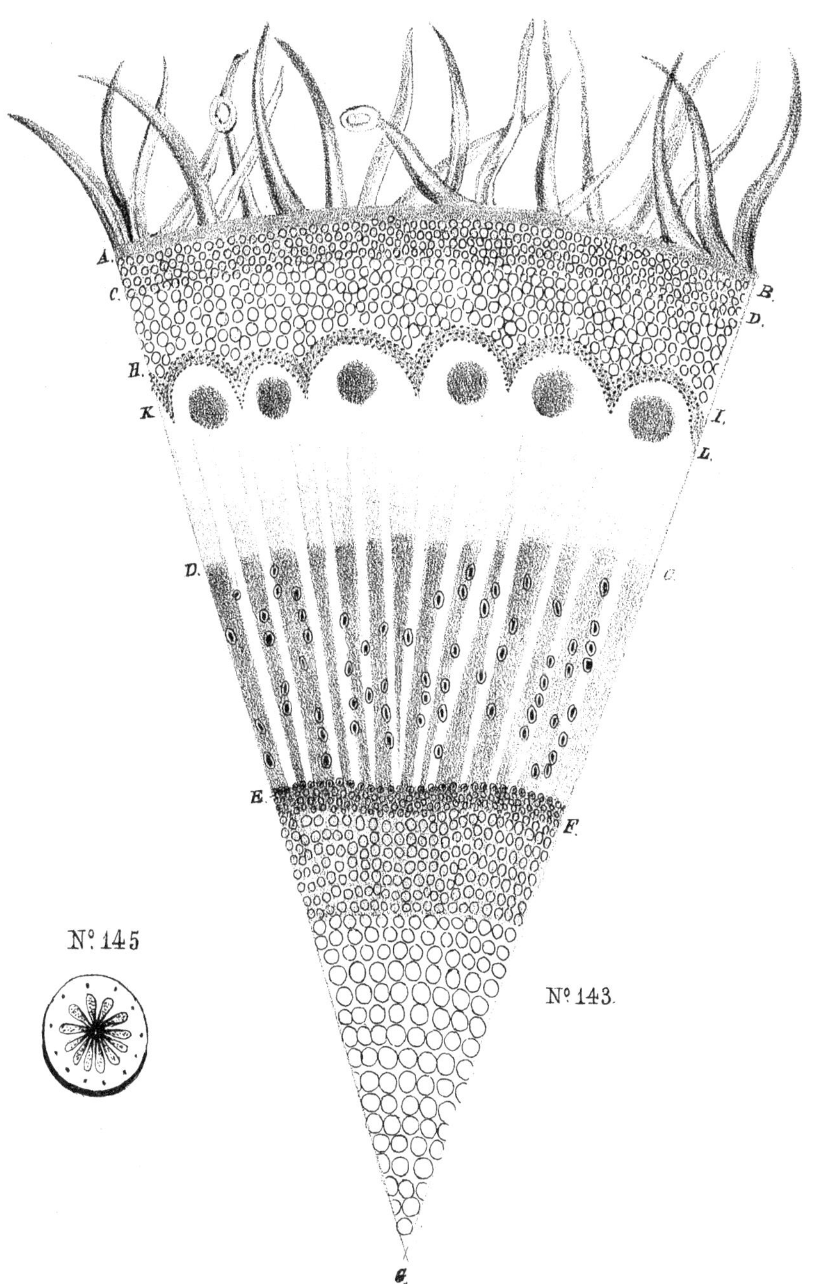

them taper to a point, but some of them, as is seen, are rounded and knobbed at the end. The narrow boundary extending from A to B, indicates the position of the skin of the bark, and within the bark itself the vessels are variously arranged. In the band just below A B they are small, much crowded together, and very compact; while those next in order towards the centre are larger. Next succeeds a row of vessels in arched clusters, extending from H to I, the cells being exceedingly small and crowded together by hundreds in one arch. Below these a ring of large tubes are seen stretching over from K to L. This latter class are termed *milk-vessels*, on account of their containing a milky liquid peculiar to the Sumach. The wood below D C is filled with pores, which seem to be disposed without regard to any particular order; but the radial divisions of cellular tissue evidently tend towards a regular arrangement. A waving band of sap-vessels extends from E to F, bordering the edge of the wood where the pith commences.

WORMWOOD.—A transverse section of a stalk of Wormwood exhibits a structure of extreme regularity, the great size of the pith, compared with that of the wood, showing at once its herbaceous character. The bark includes nearly a third of the surface, the wood occupying a smaller space, while the pith comprises all the rest of the surface. The spherical cavities of the cellular tissue form a broad ring, and within this space a number of large vessels, are situated; arranged in a circular row along the inner margin of the ring of cellular tissue. These are termed the resiniferous or gum-vessels, which secrete the aromatic fluid peculiar to the plant.

Some vessels of this kind are also found within the pith. The semi-circular figures, clusters of sap-vessels, span the sections in a row. Within the woody part the spiral vessels are seen, but quite thinly and irregularly scattered. The broad insertions in the woody part, and which diverge from each other as if proceeding from the centre, are the rays of cellular tissue, which in this plant are seen to be of comparatively great thickness, and commence and terminate in a different manner from the same rays in wood. For here, instead of being distinct lines, they are beheld arched at both extremities and united with each other. Moreover they do not terminate where the wood ends, but extend nearly half their length into the bark, enclosing the semi-circular clusters of sap-vessels. The pith, as is evident, is very porous, consisting of a vast number of large cells.

ROOT OF WORMWOOD.—The structure of the roots of plants is similar to that of the trunk, being formed of the same textures disposed in a corresponding manner.

Sections of roots display a symmetry and elegance of arrangement by no means inferior to that revealed in transverse slices of wood. In drawing 145 is delineated a cross section of the root of Wormwood of its natural size. When magnified, large circular spots are seen interspersed through the bark, which are the gum vessels of the Wormwood, which are likewise found in the bark of the stalk. Radiating from the centre of the section, and dividing it into symmetrical portions, three complete figures are seen, shaped like the sticks of an ivory fan, traversing the wood and extending into the bark. These figures terminate in the bark in clusters of vessels, through which flows a limpid fluid or sap; and within these clusters one or more gum-vessels exist. The rest of the magnified image consists of the woody portion of the root, which consists of two parts; namely, the true wood forming the lower part of the radial figure just described; and the cellular tissue, interposed between them, and running from the bark to the very centre of the root. Throughout the true wood, spiral vessels are scattered which increase in size from the centre outward. A section of the common thistle displays great beauty in the formation of the cavities of the cellular tissue. The pith consists of cells of different sizes, those of the largest kind being one hundred times greater than those in the Oak. These cells are not spherical, but are angular cavities of a regular shape, the sides of which are formed of fibres running, in most cases, horizontally and winding in a circular manner out of one cell into another; a single ring of fibre passing into no less than six cells, and constituting a side in each. Large spiral vessels are distributed throughout the woody part, which is separated into regular oval-shaped compartments by thick divisions of cellular tissue, that penetrate far into the bark; while two sets of vessels, the one filled with a milky and the other with a limpid fluid, are arranged on the outer verge of the pith in a double row of crescent-shaped clusters.

In view of the facts just adduced, we see at once the high utility of the microscope in revealing to us the true nature of the structure of bodies. Pores or vacant spaces are found diffused through the mass of bodies to such an extent, that *porosity* is one of the leading mechanical properties of matter; but in the denser bodies the pores cannot be distinguished by the naked eye. And the microscope is needed to render them clearly visible. Beneath its revealing glasses, substances which before appeared solid, are now seen, perforated with innumerable cells, which in the case of woods, occupy, for the most part, more space than their intersecting sides; and even the apparently solid sides of the larger cells yield to the higher magnifying powers, and display a porous structure.

FOSSIL WOODS AND PLANTS.

Not only is the microscope eminently serviceable to the botanist, in revealing the curious structures of living plants and their interior organization, but it is highly useful to the geologist, who is enabled by its aid to read with the utmost precision many portions of the ancient vegetable history of our globe, legibly imprinted in its fossil woods and plants. So perfectly has their structure been preserved for ages, that the skilful observer easily detects the various species, and assigns them their appropriate place in the vegetable kingdom. The substance of these woods, as Mantell remarks, is completely permeated by mineral matter. It may be lime, flint, iron, or iron united with sulphur; and yet both the external character and internal structure be preserved. Such are the fossil trees of the Isle of Portland, where a whole forest of Pines seems to have been transformed into stone, on the very spot where they grew and flourished: the roots of the trunks changed into flint, piercing deep into the soil whence they sprung. Fragments of these trees so closely resemble decayed wood, that a person who bestows upon them only a casual glance is completely deceived; but, by close examination of their texture and substance, he finds that they possess the weight and hardness of stone. In wood petrified by flint the most delicate tissues of the original remain uninjured, and are displayed under the microscope in the most beautiful and distinct manner. Wood petrified by lime also retains its structure, and in many limestones leaves and seed-vessels are faithfully preserved.

In the Egyptian and Lybian deserts, a numerous assemblage of trees has been discovered, petrified by flint. Fragments are found everywhere scattered over this arid region, but the most interesting locality is a table-land, about seven miles south-east of Cairo, where the trees are found in such numbers that it is termed the *Petrified Forest.* Here huge trunks of flint are seen crossing each other in every direction, as if swept down by the irresistible force of a hurricane.

Two of the largest, the dimensions of which were taken by Col. Head, who visited this spot, measured respectively *forty-eight* and *sixty feet* in *length*, and *two and a half* and *three feet* in *diameter* at the base. In the rich specimens collected by him from this locality, the most delicate cells and veins of the interior structure of the wood are filled with chalcedony and jasper, and some of the vessels, injected with flint of a bright vermilion and blue color, traverse the cellular tissue, which gleams with a golden hue.

Not only on the surface of the ground are petrified trees discovered, but they have been brought to the light from a depth of more than one hundred feet; where, notwithstanding they had been buried for ages, their structure was so perfect, that the species to which they belonged was at once identified. To effect this result a transverse or longitudinal section of the fossil specimen to be examined is obtained, which, after being cemented to a slip of glass, with Canadian balsam, is ground down with emery, until it becomes sufficiently thin for its structure to be perceived under the microscope. When the section is thus properly prepared, and magnified from one to four hundred times, the peculiarities in the structure of the wood are revealed with great distinctness.

Four specimens of fossil woods are delineated in figures 147, 148, 149, 150; and by comparing them with the figures 137, 138, and 139, their identity with the coniferous woods is at once perceived. Figure 147 is a transverse section of a species of Pine, petrified by flint, and taken from a quarry near Maidstone in Kent. It is magnified in length *one hundred and twenty* times, and is exhibited as it appeared when viewed by reflected light; the lighter portions representing the delicate web of fibres constituting the wood. A longitudinal section of the same wood is shown in figure 148, magnified linearly *two hundred and fifty times.* Several rows of parallel vessels are revealed running in the direction of the trunk, and each, like the White Pine, is studded along the sides with single rows of disks or glands. Figure 149 is a transverse section of coniferous wood, petrified by lime. It is magnified *eighty times*, and exhibits very clearly the cross sections of numerous rows of transverse vessels. Figure 150 is a longitudinal section of the same, magnified *one hundred and twenty times*, and shows with great distinctness the coniferous nature of the wood, for the double rows of disks alternating with each other, are seen embossing the whole range of the parallel vessels. Ferns of great beauty are preserved by petrifaction in the same manner. In the vicinity of Chemnitz in Saxony, ferns petrified by flint are found, their external surface possessing a woody appearance of a reddish brown hue, while the interior structure is of a dull red, variegated with blue and yellow, arising from the agate and chalcedony which occupies the most minute ramifications of the vessels of the plant. When slices of the fossil are ground down very thin, the microscope reveals the peculiar structure of the plant, though unnumbered years have elapsed since it was living, with as much faithfulness as the organization of the specimen which has just been gathered from the fields.

Fig. 147.

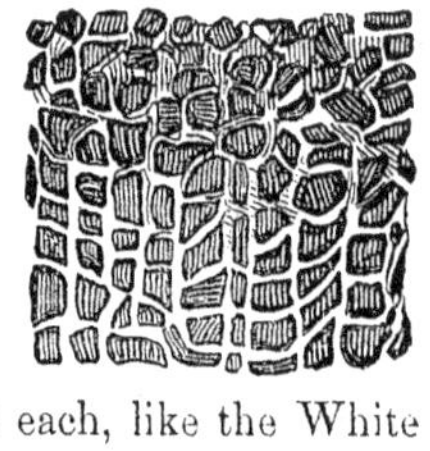

Fig. 148.

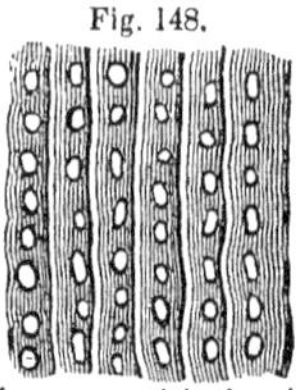

Fig. 149.

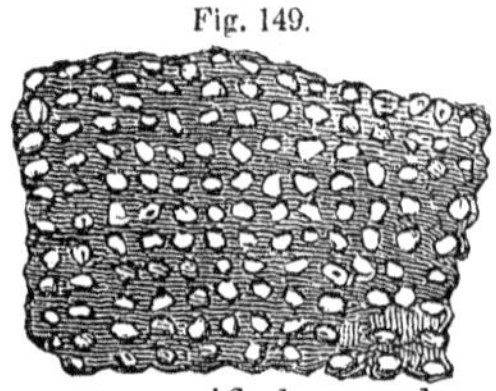

Fig. 150.

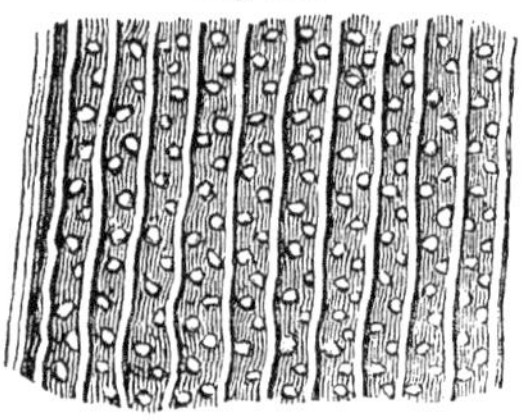

Not only are the more solid and durable portions of wood and vegetables preserved for ages by petrifaction, but the *pollen* of cone-bearing trees, like the Pine, has been found in a fossil state. In Egra, in Bohemia, a deposit has been discovered *two miles long* and *twenty-eight feet thick*, entirely composed of *fossil animalcules* and *pollen;* the first ten feet being marl filled with Infusoria, and the remaining eighteen, *pollen* mingled with fossil animalcules.

COAL.—It has been proved beyond a doubt that the vast stores of coal, which have been provided for the use of man, are of vegetable origin; and the microscope has been of essential use in enabling the investigator to detect the peculiar structure of the plants that compose it. Mantell thus remarks:—"The slaty coal generally preserves traces of the cellular tissue and spiral vessels; and dotted cells, indicating the coniferous structure, may readily be detected, by the aid of the microscope, in chips or slices prepared in a proper manner. In many examples the cells are filled with an amber-colored, resinous substance; in others the organization is so well preserved, that on the surface, exposed by cracking from heat, cellular tissue, spiral vessels, and cells studded with glands, may be detected. Even in the white ashes left after the combustion of coal, traces of the spiral vessels are discernible, with a high magnifying power. Some beds of coal appear to be wholly composed of minute leaves; for if a mass be recently extracted from the mine and split asunder, the exposed surfaces are found covered with delicate pellicles of carbonized leaves and fibres matted together, and flake after flake may be peeled through a thickness of many inches, and the same structure be still apparent. Rarely are any large trunks and branches observable in the coal, but the appearance is that of an immense deposit of delicate foliage."

CHAPTER V.

CRYSTALLIZATIONS.

"The crystal drops
Shoot into pillars of pellucid length,
In forms so various, that no powers of art,
The pencil or the pen, may trace the scene.
Here glittering turrets rise, upbearing high
Large growth of what may seem the sparkling trees
And shrubs of fairy-land. And fretted wild,
The growing wonder takes a thousand shapes
Capricious." COWPER.

ONE of the most beautiful discoveries of science is that which reveals the singular fact, that when bodies pass from the liquid to the solid state, with a proper degree of slowness, they assume forms peculiar to themselves, which are often characterized by great elegance and beauty. These configurations are termed *crystallizations*, and each crystalline substance is regarded as having an original form, called the *primitive crystal;* a number of which, combining in various ways, frequently give rise to a rich assemblage of the most exquisite and symmetrical figures.

The greater part of the solid bodies that compose the mineral crust of the globe are discovered in a crystallized state. This is true, for instance, of granite, which consists of crystals of quartz, feldspar, and mica; and vast hilly ranges of clay-slate are likewise constituted of a multitude of regular forms. The body before crystallization may exist in the fluid state, either from combining with a liquid, or from the action of fire. Brine is an instance of the first condition. Here the salt is thoroughly dissolved, so that a particle cannot possibly be seen; but if the solution is slowly evaporated, the salt again appears in the form of cubes. An example of the second mode of action is afforded in the case of sulphur, which, when melted and suffered to cool gradually, shoots out into crystals, which, if undisturbed, are soon blended into a compact mass. The original atoms are so inconceivably small, that not only do they escape the unaided eye, but even when it is assisted by the most powerful glasses they still elude its utmost range. Nevertheless, when, in the act of crystallization, particle begins to unite with particle, the microscope is of great utility, during the earlier stages of the process, and crystals of the richest configuration are then seen forming immediately under the eye, branching in every direction, with the most wonderful regularity and symmetry; and often bearing a striking resemblance to the most beautiful and graceful foliage, or crowding together in glittering star-like clusters. It is only those substances which crystallize rapidly whose beautiful forming figures can be seen by the microscope, and in order to render them visible, the following process is employed :—The salts are first dissolved in water until the liquid is thoroughly saturated, and then, when it is desired to view the crystallization, a drop or two of the solution is spread over a clear strip of glass, which has been previously warmed. The watery film is then placed near the focus of the object-glass of the microscope, and if the solar microscope is employed, a large image of the liquid on the glass is seen upon the screen. As soon as the fluid is sufficiently evaporated, the dissolved salt is beheld changing rapidly from the fluid to the solid state, and branching over the whole screen in crystals of the most exquisite forms ; a single crystal, in certain cases, often apparently shooting the length of *six or eight feet*, in the course of half a minute. In the present chapter we shall describe some of the most interesting crystallizations.

Nitrate of Potash, or Saltpetre.—When this salt is dissolved in water, and a few drops thinly spread over a glass slide, crystals are beheld shooting inward from the edges of the fluid, upon the application of a gentle heat. The crystals are very transparent, and their primitive form is that of six-sided prisms. In drawing 151 many varieties of crystals are delineated, as they appear under the microscope in a crystallized film, moderately magnified. If the crystals form with great rapidity, long arrow-headed shafts, like that portrayed at C, are seen shooting swiftly along and throwing out lateral spurs from one side, forming a figure like B. These lateral branches run parallel to each other, and from their sides secondary branches likewise emanate, spreading over the surface in lines of crystal network. If the process of crystallization advances with less freedom, the lateral branches are not formed ; but the main shoot appears arrow-headed, with jagged sides, as in the figure C ; the sides or teeth being the rudiments of the lateral spurs. In the field of view other forms are seen like those in the group D, most of which resemble the variety C in their incipient formation. Another kind with regular faces is seen at A ; the breadth of this crystal at the end A, as measured by the micrometer, was found to be *one-two-hundred and seventy-seventh part of an inch.*

In India, saltpetre forms upon the surface of the ground in silky tufts and slender prismatic crystals ; especially when the abundant rains of this tropical region are succeeded by hot weather. These delicate filaments are swept from the surface of the soil into large heaps, which are then leached like ashes, and the liquid thus obtained, after being suffered to settle, is evaporated, when the nitre remains in a crystallized form.

In certains regions of India the lower part of the mud-walls of the houses becomes wet and black each morning during the dry season, from February to May ; and portions of the mud crumble down into a fine powder. This dust is swept up every day, and contains about one-fifth of its weight of saltpetre. It is stated by the natives that the supply is abundant during those years when the preceding monsoon-storms have been most heavy, and the thunder and lightning that attend them unusually frequent. The earth from which the nitre has been extracted, in a year or two becomes impregnated again, and the tendency of the soil to reproduce it causes much trouble and annoyance to the occupants of houses. Bishop Heber remarks "that the nitre can scarcely be prevented from encroaching, in a few years, on the walls and floors of all lower rooms, so as to render them unwholesome, and eventually uninhabitable." To such an extent does it prevail at Tirhoot, that it may be brushed from off the lime walls of the houses, and other humid places, almost in *basketfuls*, every two or three days.

FLOWERS OF BENZOIN.—A species of gum, known by the name of *Benzoin*, is extracted from a tree which grows in Java, and some other parts of the East. An incision is made into its trunk and branches. and a fluid exudes from them, which hardens upon exposure to the air, concreting into brittle masses. It melts when subjected to a moderate heat, and sends forth a thick, white smoke, which condenses, upon the underside of the cover of the vessel containing the melted gum, in slender and delicate crystals of benzoic acid. These are beautifully white and transparent, and emit a fragrant odor. A drop of the solution of this acid exhibits very elegant crystallizations under the microscope. Sharp crystals are first perceived forming at the edges, transparent, and without color, which soon push forward towards the centre of the drop, in the form of running vines and beautiful tufts of mimic foliage. Several specimens are delineated in drawing 152, possessing the same characteristic form, but still differing in some particulars. All consist of similar minute crystals gracefully clustered together; but while one shoots along in light and airy tracery, as in the right hand figure, another, like that upon the left, extends laterally, and spreads its glittering branches from side to side; and in different parts of the field of view other configurations start forth, and rich tuft-like figures are seen like the central forms of the group. The largest tufts and vines appear dark to the eye from the immense number of minute crystals which are there clustered together; but amid these, under a subdued light, wreaths of exquisitely delicate foliage are seen, formed of the purest crystals, and gleaming like silver sprays. Interspersed with the rest, crystallizations in the form of crosses occur, as shown in the drawing. When the acid is dissolved in alcohol, and the solution spread upon the glass, the crystallization proceeds with great swiftness, on account of the rapid evaporation of the spirit. At one moment the eye of the observer gazes upon nothing but a film of liquid, and at the next, on a sudden, at a single flash, order springs forth, and the chaotic surface is profusely studded with all the exquisite and graceful combinations which have been detailed. The delineations in the figure are drawn from actual crystallizations, like all the rest, and represent forms of average dimensions. Some idea may be gained of the smallness of the crystals, from the fact, that the breadth of the group *a*, *b*, is only the *one-six-hundred and twentieth part of an inch.*

PLATE VI.

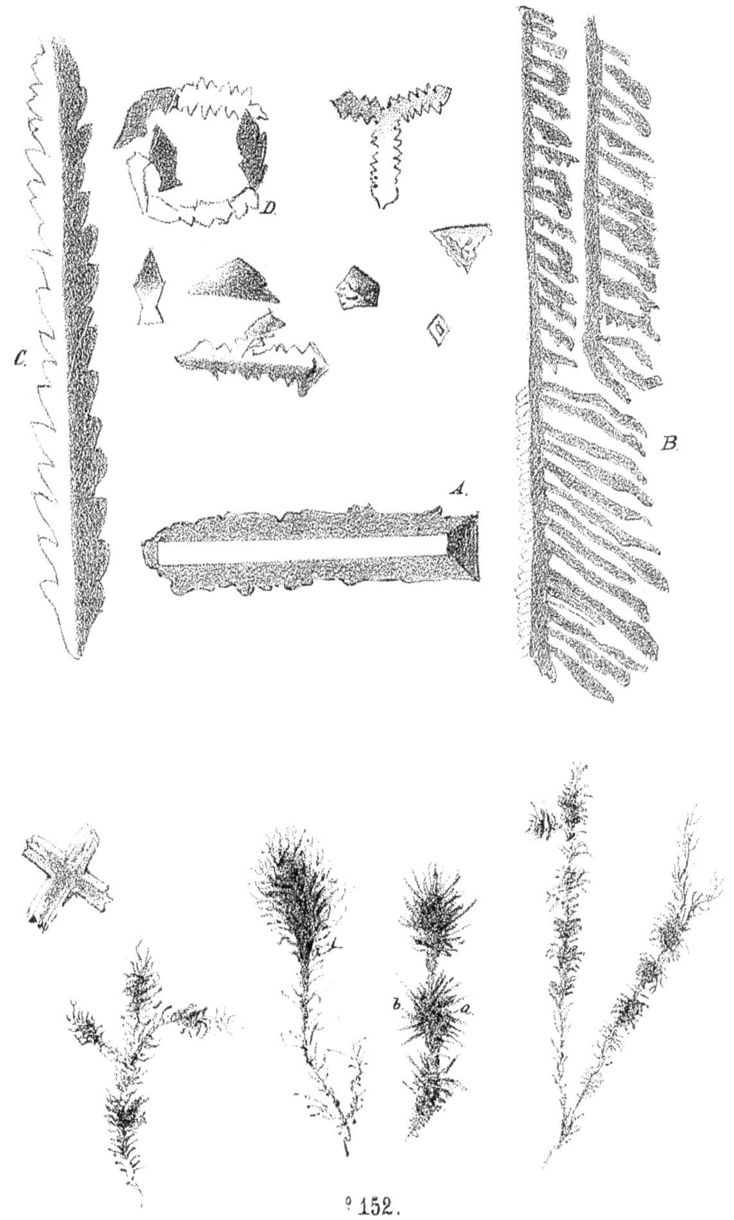

Sulphate of Iron, or Copperas.—This substance crystallizes in transparent rhomboidal prisms, and appears of a sea-green color when the crystals possess a considerable size. Under the microscope it displays very regular and interesting combinations. A drop of an aqueous solution of the sulphate of iron must be only moderately heated, when the film of liquid is soon perceived, crystallizing at the edges where it is thinnest; the principal crystals pushing forward in a straight direction, while at the same time branches proceed from them on either side. These lateral shoots all start from the main crystal at the same inclination, and advance parallel to each other with the greatest precision and order, throwing out likewise secondary branches, which meet and combine; and the whole array of interlocked crystals, its line bristling with arrow-headed forms, is seen steadily advancing over the field of view. Some of these configurations are massive in their structure, and others more light and delicate; but all more or less reveal the form of the primitive crystal; and minute as they are, their solidity is apparent from the mingled lights and shadows that fall upon the crystallized surface. And very beautifully are these lights and shadows varied, as the mirror is differently adjusted, and the illumination now plays brightly upon some rich and glittering cluster, and again falls chastened and subdued upon the mimic gems. Minute crystals, possessing the primitive rhomboidal figure, are sometimes found at the edges of the crystallized film, often clustered together in grotesque combinations, resembling, with their salient points and re-entering angles, the frowning bastions of a fortress.

Camphor.—When camphor is dissolved in alcohol, very elegant crystals are formed upon a slip of glass, by spreading, in the usual manner, a drop of the solution over the surface.

The film of the fluid crystallizes with great rapidity, owing to the rapid evaporation of the alcohol. When the glass is just prepared and placed under the solar microscope, the image of the drop is beheld upon the screen, as a uniformly misty surface; suddenly it is broken up in the thinnest part, which in a moment is studded with beautiful star-like figures. Instantaneous flashes now flit successively over the remaining portions of the cloudy field, and simultaneous with this motion, the same elegant radiated figures start forth, perfect in form on the ground over which passes the creative wave. These configurations are extremely beautiful, and consist of delicate crystals, which radiate from a centre, most commonly in six branches, which are nearly of the same length. They are formed like fern-leaves, wide at the base, and gradually tapering to a point, the fringes at the sides being composed of slender, feathery crystals. Scarcely any heat is necessary to produce these configurations, for the spirit quickly evaporates, and the crystals that originate are of short duration, in consequence of the camphor itself being volatile. The crystals of camphor are very minute, the entire length of a branch being only the *one hundred and twenty-fifth part of an inch.*

Sal Ammoniac, or Muriate of Ammonia.—The crystals of this salt are among the most elegant of those which the microscope reveals, and the prepared solution crystallizes with great facility. The change commences at the edges of the film, and at those places on the surface where the liquid is thinnest; and from these points sharp, broad, dagger-shaped crystals push out in all directions. The crystal appears at first as a single stem of the most perfect transparency, but as it advances it throws out at each side blunt crystals of different lengths, parallel to each other, and usually at right angles to the main shoot. These lateral spurs increase in length until the middle of the principal crystal, when they gradually diminish in extent until they vanish at its remote extremity. The lateral branches often extend from the principal crystal to a considerable distance, and are themselves studded with minute crystals at the side, which also shoot out at right angles; and from these again similar systems proceed to an indefinite extent. Another combination consists of six broad and beautiful leaves diverging from a common centre. In some of the specimens the leaves are without branches; but in the others they break forth into crystals on either side; and each of the six stems becomes a silvery spray. A very common form is a transparent, dagger-shaped crystal, with the blade, handle, and guard all complete. The crystals are quite small, the breadth of the main stem, in some, being only *one five-hundredth part of an inch;* and yet, small as they are, these minute forms exhibit with distinctness, when the light falls upon them in a proper direction, the full, rich, and vivid play of the prismatic colors.

Muriate of Barytes.—When the muriate of barytes is dissolved in water, it forms a clear and colorless solution, which speedily crystallizes on the glass slide by the application of a very moderate heat. The resulting configurations are of exceeding beauty, and no verbal description, or delineation of the artist, can convey to the mind a full conception of the richness and elegance of the forms that are presented the eye by the magic power of the microscope. Not only is the beholder charmed with the wonderful delicacy and exquisite grace of the figures; but such is the swiftness with which the fluid crystallizes, that he sees them in the very process of formation, darting forth their glittering filaments in all directions with a velocity truly astonishing. A quick formation belongs to nearly all the crystallizations herein described; but the very rapid change of the muriate of barytes from the liquid to the fluid state peculiarly impresses this circumstance upon the mind. One combination common to this salt is found near the edge of the crystallized field. It appears like a collection of shrubs. shorn of their leaves, growing up from the midst of a tuft of rank herbage. The main crystals take no particular direction in reference to each other, and the lateral branches appear likewise to be guided in their course by no especial law. From crystals like these numerous branches proceed, dividing and sub-dividing until an infinity of boughs and sprays are seen, rising from a single stem. and groups and groves of crystal trees spread their fairy foliage over the whole field of view.

One part of this last assemblage of crystals is singularly connected with the rest; for on the same side, from a single point in the principal crystal, two shoots emanate obliquely from it and at right angles to each other; but the lateral spurs from these observe the same laws, in regard to direction, as those in other parts of this combined figure. The size of the crystalline stems is exceedingly small, the breadth in ordinary specimens being only *one-seventeen hundred and sixtieth part of an inch.*

The crystallizations of the muriate of barytes, like many others, exhibit an extremely beautiful appearance when viewed, not by the diffuse light of day, but by a single light, as a lamp. Each crystal then acts as a prism in decomposing the rays, and the entire field of view becomes illuminated with the splendor of the sevenfold tints of the rainbow.

In a third variety the main crystals are short and thick, their ramifications occupying only a little space. The secondary crystals are parallel to each other and perpendicular to the parent stem; and from these a third system proceeds, governed by similar laws.

Another beautiful configuration is sometimes formed in which the crystalline stems radiate from a common centre, and diverging more and more as they recede from this point, push forth on either side buds and shoots of sparkling crystals, covering the entire circle throughout which they extend with clustering gems.

Bichromate of Potassa.—This salt produces very elegant combinations, the original form of the crystals being that of a four-sided prism. The solution is of a transparent cherry color, and the minute crystals seen by the microscope gleam with a rich amber light. Like those of the muriate of ammonia they form with great rapidity, and the swift advance of their spreading configurations gives full employment to the eye of the observer. A represents the primitive crystals as seen at the edges of the film, but the two most common forms are running vines and plume-like tufts, consisting of numerous crystallized branches of the most delicate structure. The first is often seen originating in a single stem, which, as it grows, soon breaks up into a thousand curved shoots, that interlace and entwine with each other, composing a kind of irregular crystallized network extending over the surface before occupied by the liquid. The lateral spurs are short, forked, and disposed along the stem without any particular regard to symmetry; and are often loaded with comparatively heavy crystals, the whole presenting an appearance not unlike a spray of withered herbage fringed with crystals of hoar-frost. The second kind has numerous slender ramifications radiating from a single stem, each filament being studded at the side with minute crystals.

These glittering plumes are scattered in profusion over the whole field of view, amid the sparkling network of crystallized vines, and the union of these rich and radiant configurations dazzles the eye with visions of rare and surpassing beauty.

A singular form is sometimes presented, in which the solution crystallizes in circles around a particle of sediment; the circles are gemmed with crystals at the sides, and terminate in branching sprigs as graceful as the leaflets of a flower. These delicate sprays are very small, the breadth of the crystals measuring only *one-sixteen hundred and sixtieth part of an inch.*

SULPHATE OF SODA, OR GLAUBER SALTS.—This salt crystallizes slowly by the application of a gentle heat, and exhibits a great diversity of combinations, which are, for the most part, massive, and stand boldly out upon the surface of the glass. One variety commences in a spicular cluster, similar to that delineated at A in drawing 158, the branches of which spread out in long needle-shaped crystals on every side, which, intersecting with others of similar nature, frequently form an irregular crystallized lattice work. Sometimes long and massive crystals radiate from a common centre like the spokes of a wheel. Another variety of crystal, of a delicate white color, broad, pointed, and shaped like a feather, is often seen advancing in the field of view, and sending forth its glittering filaments on either hand. In other parts of the crystallized film, a number of these crystals are beheld ranked side by side, like the teeth of a comb, and the surface of each is itself studded with still smaller crystals. Rich, starry crystals are also found, like those displayed in group B, and the other forms which are here delineated are scattered in profusion amongst the rest. Very beautiful figures are frequently observed near the edges of the drop where the salt is most abundant.

Two of these are exhibited at C and D, the first of which is a heavy transparent configuration of considerable thickness with serrated sides, formed of single diamond-shaped crystals. The second is a very singular crystalline structure, and

PLATE VII

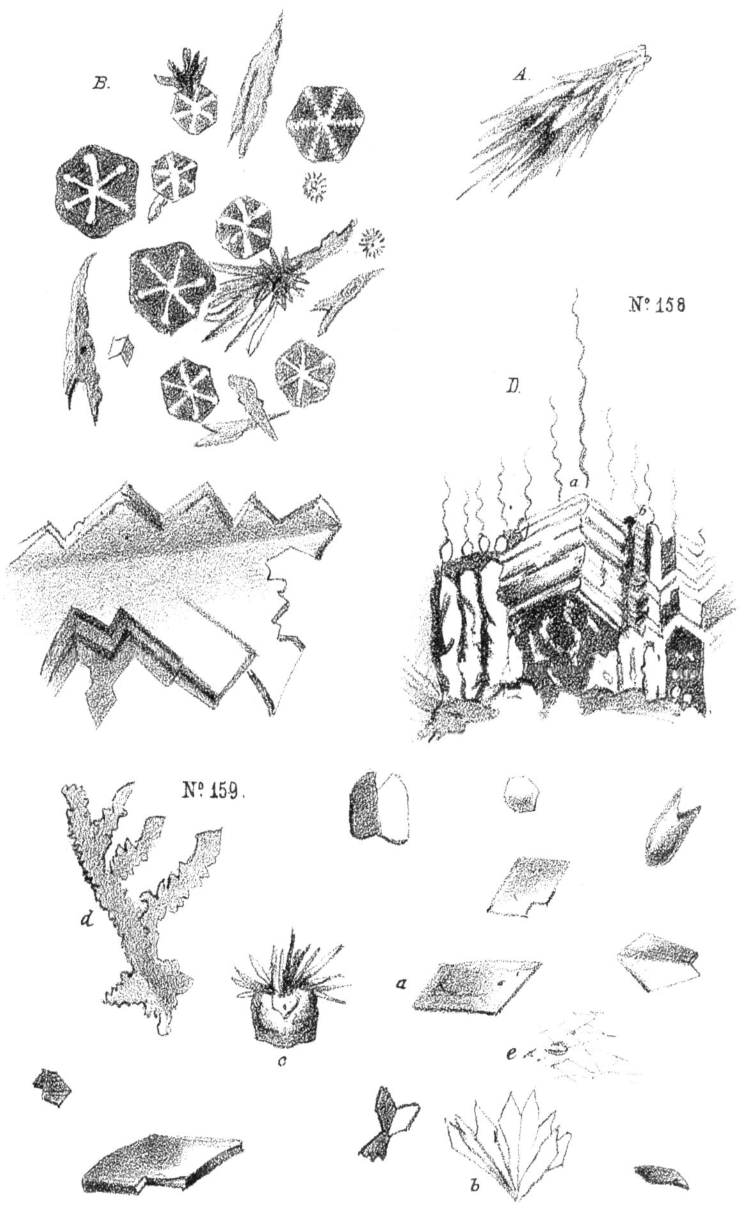

resembles some massive piece of sculpture wrought with elaborate skill. The crystals that compose the arch-like figure are very large in comparison with the rest in this cluster, and yet the distance from *a* to *b* measures only *one-three-hundred and fifty-seventh part of an inch.*

VERDIGRIS.—The crystals of verdigris are of a fine greenish blue color, and have for their primitive form that of a lozenge or rhomboid. When a drop of the prepared solution is placed upon a slip of glass, it begins to crystallize at the edges, under the action of a mild heat, and clusters of transparent crystals, of the form exhibited at *a*, in drawing 159, are seen gleaming with a rich blue tint, upon the edge of the drop. Another form like that at *b* is likewise beheld branching from a single stem, like the leaves of the fleur-de-lis. A similar spicular cluster, in which the stems are more numerous and slender, is observed at *c*. Another configuration is delineated at *d*, which originates in a single, diamond-shaped crystal, one point of which, possessing greater energy than the others, has pushed forth a long serrated crystal, from which arrow-headed lateral forms proceed, in a direction parallel to each other.

The solution of verdigris does not crystallize with rapidity, and as the liquid evaporates slowly in the central portions of the drop, delicate needle-shaped crystals are detected amid the larger forms, crossing each other in all directions. A specimen of this configuration is drawn at *e*, and throughout the entire surface of the crystallized film, minute crystals of the first form are profusely scattered. The larger crystals of verdigris are extremely well defined, and as perfect as if cut by the lapidary. In the figure *d* the breadth of one of the lateral crystals at the head *e f*, is *one-five-hundredth of an inch.*

SULPHATE OF MAGNESIA, OR EPSOM SALTS.—The crystals of this salt combine in figures of exceeding beauty, and with a slow and steady motion. The observer is thus enabled to examine at full leisure the shapeless fluid, as it gradually changes into the richest configurations, which grow and expand on every side, lavishly adorned with the most exquisite and singular figures. The crystals are best viewed in the evening, by the light of a lamp; at the moment they begin to form, they are then seen shooting along parallel to each other, in the shape of massive broad shafts, arrow-headed, and serrated on either side, At times their structure is more elaborate, and they somewhat resemble long leaves, with strongly marked veins branching from the main stem. As the crystals advance, the lateral points or teeth likewise expand in broad crystals, running, some obliquely and some at right angles to the principal figure. The lateral crystals likewise throw out from their sides a third set, and thus the ramifications extend until, weaving and interlocking with each other, the entire field of view is covered with the crystalline structure. This in many parts resembles, in the promiscuous grouping of its figures, a surface composed of fern leaves placed upon each other, without any regard to regularity; but in others the figures are seen interlacing in the most fantastic shapes. A remarkable characteristic of these crystals is their softness of tint; they shine with a pearly whiteness, and the light comes through them mild and subdued, like the gentle radiance of the moon, imparting such a softness to these beautiful figures that they appear as delicate as flower-wreaths, wrought on a satin ground. Another form is a pearl-colored, lozenge-shaped crystal, and is often seen small at first, but of perfect proportions. As the eye remains fixed upon it, it is seen gradually to increase in size, retaining nevertheless, at the same time, its original symmetry, a result which is effected by the sides expanding uniformly from the centre of the crystal. Several modifications of this type are seen. A fourth variety consists of clusters of spicular crystals, and in another configuration massive crystals cross each other, and interlock like the roots of trees.

SULPHATE OF COPPER.—The sulphate of copper affords a fine blue solution, which forms quickly, upon the application of a moderate heat, into long spicular crystals, uniting and blending with each other. The network of the crystals is clearly discerned by the unaided eye, but the microscope is needed to display their more intimate combinations, and the different configurations they assume.

In drawing 161 are delineated a number of crystals of sulphate of copper; the long spicular figure C, at the top, is that which is first detected under the microscope, pushing forward its sharp point into the crystallizing fluid. As it advances it spreads out laterally, sending forth numerous crystals, mostly from one side, which often unite so compactly together, as to constitute a triangular plate. The form it then assumes is the same as that which is exhibited at D.

A more delicate and perfect configuration is displayed at *a b*; the crystals of which are exquisitely fine, especially those at the sides. Instead of being blended together, as at D, they are distinct and separate from each other; and in every set branch from the main stem in exactly parallel directions. The distance between the two adjacent principal crystals *a* and *b*, is only the *one hundred and ninetieth* part of an inch. The cluster E consists of short prismatic crystals, which are usually found in groups, near the edges of the liquid, wherever the film is comparatively thick, and the matter held in solution quite abundant.

Profusely scattered throughout this locality, fine diamond-shaped crystals are likewise often found. A rare configuration is delineated at F, consisting of a minute and distinct portion of the liquid which has concreted without regularly crystallizing throughout, a circumstance which often happens when too much heat is applied to the glass slide. In the midst of this mass a beautiful spiral system of exquisitely formed crystals is beheld, emanating from a central point, and spreading its slender and plumy branches over the entire transparent surface.

PLATE VIII.

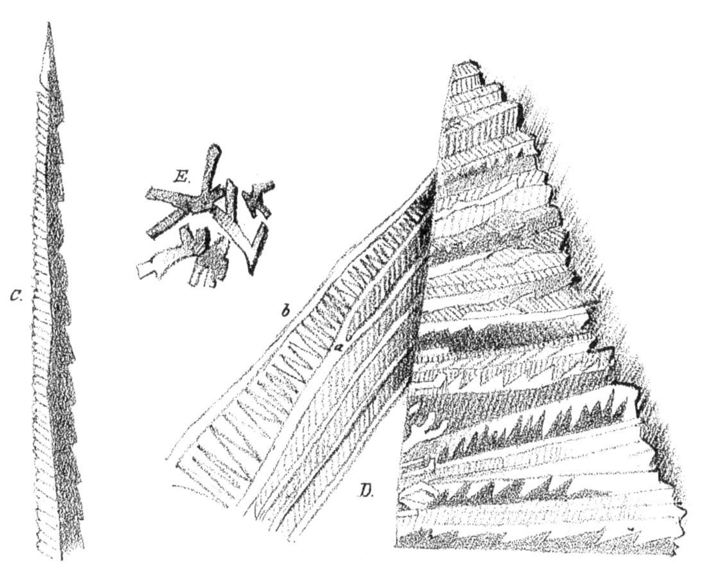

№ 161.

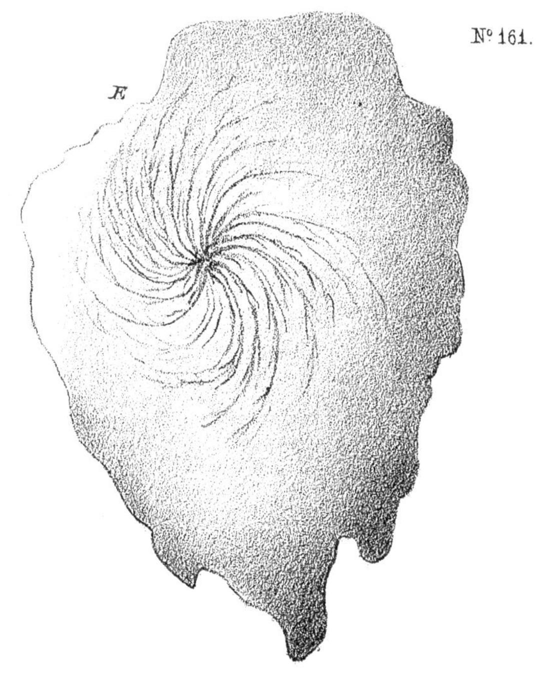

ALUM.—In drawing 162 are delineated several groups of the crystals of alum. This substance is for the most part artificially produced, and is seldom found in a native state, though it occasionally appears as an efflorescence, and exists in certain mineral waters in the East Indies. The primitive form of the crystals of alum is an octahedron; that is, a regular solid contained within eight equal faces. The combinations of the single crystals are extremely rich, and the figures in the drawing but faintly represent the exquisite crystalline tissue which is seen beneath the microscope emerging from the shapeless fluid; every part symmetrically wrought with lines of fairy gems, replete with elegance and beauty. The crystals form rapidly, and are beheld on the slide shooting forth into the liquid film, in figures like those delineated at A, being arrow-headed and serrated at the edges. Advancing side by side in parallel lines, they each spread rapidly on either hand, throwing out lateral spurs at right angles to the main crystals. From the edges of these secondary crystals, others in like manner dart forth, all weaving and interlacing with each other, and forming, with other similar systems, one unbroken, glittering sheet. The advancing line of such a crystalline field, is displayed in figures B, C, and D. Sometimes the main crystals, as at C, slightly diverge from each other. An unusual configuration is delineated at E, where two sets of crystals, bending in parallel curves, intersect each other, and form, by their union, a light and graceful gothic arch. The breadth of the larger crystals across their arrow-heads, *l. k*, measures about *one-two hundredth of an inch.*

SALT, OR CHLORIDE OF SODIUM.—The primitive form of crystals of common salt is a cube, and the crystalline structure of the fragments and masses in which it is found, is seen at a casual glance with the unaided eye. A solution of salt, as it crystallizes, is never seen spreading out into beautiful ramifications; but as fast as the fluid evaporates, the surface of the glass becomes studded all over with minute and sparkling gems of salt. Eight-sided and twelve-sided figures are likewise formed by the union of the primitive cubical crystals. Another variety which is quite common is that of a hollow rectangular pyramid. It begins its formation at the surface of the fluid with a small cube, upon the upper edges of which four rows of small cubes soon crystallize. To their upper and outer edges other cubical crystals now attach themselves, and by degrees a hollow pyramidal structure is completed—capped at the smaller extremity with the original cube.

SNOW.—The snow-flake, which varies from more than an inch to *seven-hundredths of an inch* in diameter, consists of an assemblage of exquisitely minute crystals; and from its beautiful figures and rich diversity of forms, has ever excited the admiration of observers.

Fig. 164.

1
2
3
4
5
6
7
8
9
10
11
12
13
14
15
16
17
18
19
20
21
22
23
24
25
26
27
28
29
30

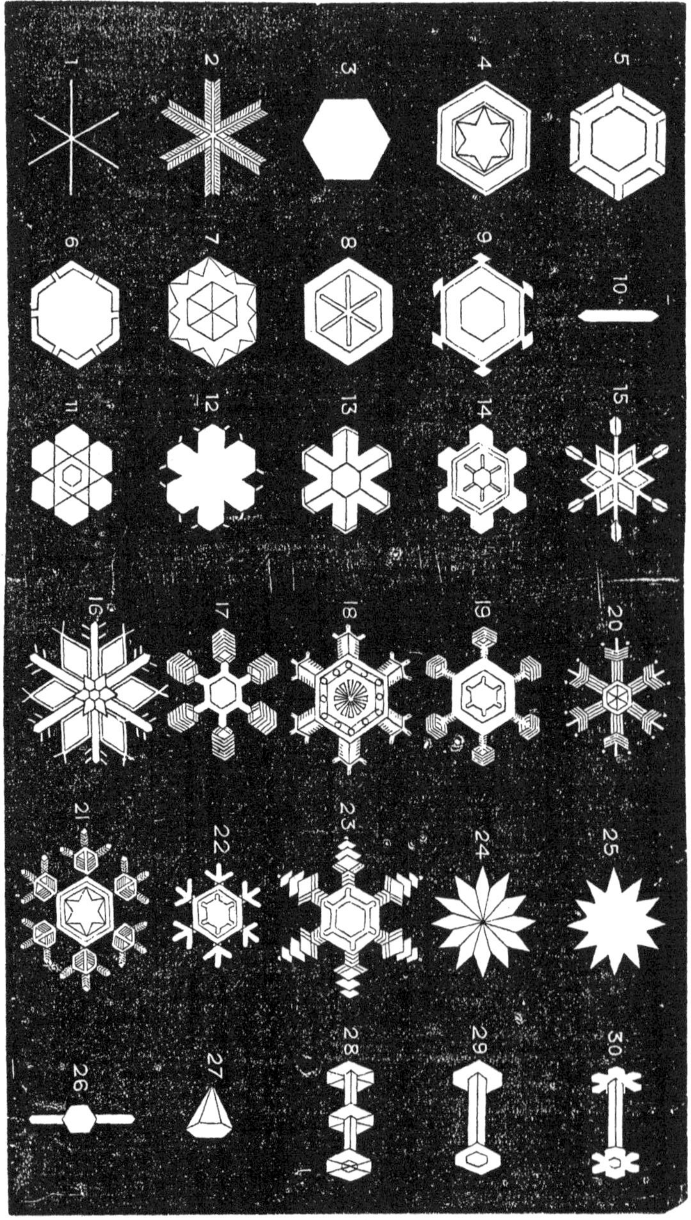

When the snow descends in a calm atmosphere, the constituent crystals of the flake are perfectly developed, but any agitation of the air, or an increase in moisture or temperature, destroys their delicate structure. The single crystals always unite at angles of *thirty, sixty, and one hundred and twenty degrees*, but by their different modes of union, give rise to several hundred distinct varieties. Scoresby, a celebrated Arctic navigator, has enumerated no less than *six hundred* kinds, and delineated ninety-six; and Kæmtz, a German meteorologist, has observed twenty more, not figured by Scoresby.

Although the varieties are so numerous, they are all comprised under *five* principal classes, which are distinguished as follows: First—crystals in the form of plates, very thin, transparent, and of a delicate structure. This class includes many remarkable varieties, which are represented by the first twenty-five forms in cut 164. Secondly—flakes either possessing a spherical nucleus, or a plane form studded with needle-shaped crystals, like the 26th figure in the cut. Thirdly—slender, prismatic crystals, usually six-sided, but sometimes having only three sides. Fourthly—pyramids with six sides, as shown in figure 27. Fifthly—prismatic crystals having, perpendicular to their length, both at the ends and in the middle, thin six-sided plates as delineated in figures 28, 29, and 30. The last two classes are extremely rare, Scoresby having observed the fifth but *twice*, and the fourth only *once* in all his voyages.

The crystallization of aqueous vapor is beautifully displayed when a thin film of moisture is frozen upon a window pane. Then, in addition to single, star-like crystals, exquisite branching configurations are seen, extending their glittering lines in all directions. When water, in a body, begins to freeze, similar results occur, and at such times, along the edge of a rivulet, long, needle-shaped crystals will be seen, darting from the ice that fringes the bank towards the centre of the stream, and which, rapidly interlacing with each other, soon unite into one compact mass. Often, upon raising a thin sheet of ice from the water, the under surface will be observed covered with a network of crystals. The snow, on account of its light and branching crystallization, descends softly upon the earth, clothing its surface with a fleecy mantle, which effectually shields the tender plants from the inclemency of the wintry season. If it had been ordered otherwise, and all the moisture, that now forms the snow, had fallen in solid masses of ice, like hail, the evils which would have arisen under such a provision in the economy of nature, must have been many and great.

ON CRYSTALS FOUND IN PLANTS.

It has been proved, by the microscopic examinations of distinguished naturalists, that saline substances are spontaneously crystallized within the cells of plants; the crystals having been found existing in infinite numbers throughout the bark, wood, and leaves of a great variety of trees and shrubs. The facts stated in this section are mostly taken from an interesting paper, read before the Association of American Geologists, by Professor J. W. Bailey, of West Point, detailing numerous discoveries made by himself. The attention of Prof. B. was accidentally led to the pursuit of this subject by noticing, one day, the ashes of a hickory ember, in which the natural structure of the wood was preserved uninjured, by the saline matter which had resisted the action of the fire. In order to preserve this structure, the Professor prepared a slip of glass with melted Canada balsam, and touching the ashes gently with the adhesive side, the delicate longitudinal section was transferred to the balsam, and became firmly fixed in this substance as it cooled and indurated; each part of the structure retaining the same relative position as it possessed in the wood. When the preparation was placed under the microscope, long rows of polygonal bodies of a brownish hue were clearly perceived. Similar bodies were discovered in the ashes of the oak, and in those of most dicotyledonous* trees, both native and foreign, constituting a large proportion of the insoluble matter of the ashes.

Prof. Bailey was at first in doubt, whether these bodies were in fact true crystals, or simply saline matter which had taken the form of the cells in which it had concreted.

This doubt was solved by observing the bark of hickory when illumined by the rays of the sun; numerous glittering particles were then seen, which proved, on examination, to be *crystals;* for when thin layers of bark, or sections of wood and bark were viewed by a microscope, the crystals were detected imbedded in their natural position.

They were, however, better seen by scraping the bark upon a plate of glass, upon moistening which with the breath, the crystals were made to adhere to the surface, while the woody particles were readily blown off. When placed under the microscope the glittering atoms then appeared as beautiful transparent crystals, having the forms exhibited in figures 165, 166, 167, 168, 169 and 170; some being single as in 165, 166, and others possessing a compound form, as shown in figure 168.

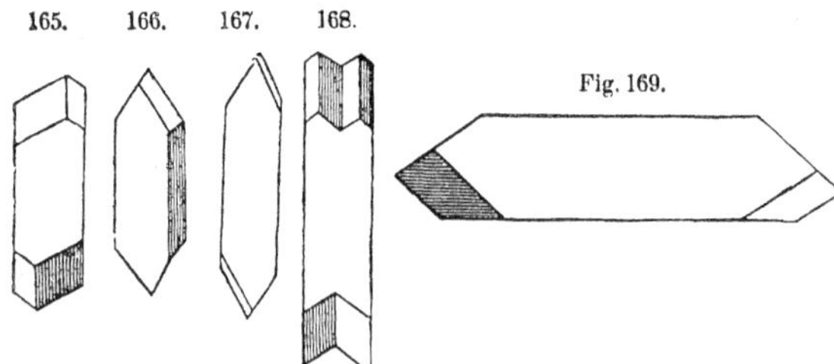

Fig. 169.

These crystals, when prepared with balsam, were identical in every particular with the polygonal bodies found in the ashes.

These singular and interesting results led Prof. B. to extend his investigations, and he had the pleasure of discovering that the bark of every species of oak, birch, chestnut, poplar, elm, locust, and of all the common fruit trees, as the apple, pear, plum, cherry; and likewise of a great number of others, were filled with crystals crowded together in vast numbers. When thin layers of the

* From the Greek dis, *double*, and cotyledon, *a seed-leaf*. Trees whose seeds divide into *two parts*, as the sprout.

bark were moistened, and examined by the microscope, the arrangement of crystals appeared like an elegant piece of mosaic work, as shown in figure 171, which is a section of the bark of a species of poplar, the crystals in the cells of the bark being either single or compound. The bark of the locust, willow, chestnut and various other trees exhibits a similar appearance. In the densest woods, such as mahogany and lignum vitæ, the crystals may be found by scraping the wood into a watch-glass filled with water, picking out the woody particles and then examining the residue; and if by this process the crystals are in any case sparingly discovered, they will be revealed in great quantities if the ashes of the wood to be examined are imbedded in balsam in the manner before described.

Fig. 171.

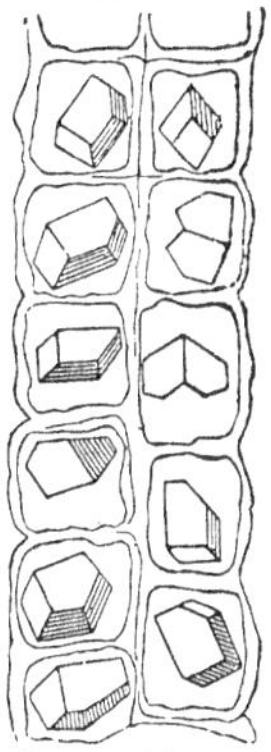

The crystals are likewise detected in the minute particles that fall from worm-eaten wood, or sawdust, and in the finer particles of ground dye-woods, such as fustic Brazil wood, camwood, logwood, sandal wood, &c.

Prof. B. next proceeded to examine the leaves of trees, which were likewise found to abound in crystals. By slowly and carefully burning the leaf until the ashes became white, and covering the residuum with Canada balsam, the incombustible portions of the leaf exhibited a skeleton of its figure. When a full grown leaf was thus prepared and placed under the microscope, the course of the minutest veins in the leaf was seen traced out in the ashes by a row of transparent crystals. In young leaves these crystals were observed only to exist in the main stem, and along some of the principal branches. In the leaves of other plants the arrangement of the crystals was found to be different, being scattered throughout the cellular tissue in star-like groups. The crystals of the primitive form, represented in figures 172, 173, 174, 175, 176 and 177, were found by Prof. B. to exist abundantly in more than one hundred species of plants, belonging to more than thirty different families, which comprise the great majority of dicotyledonous trees and shrubs, besides many herbaceous plants. The primitive form displayed in figures 165, 166, 167, 168, 169, 170 and 171, was found to be far more sparsely scattered in dicotyledonous plants, than the forms found in the last set of figures; while a third form which is exhibited in figures 178 and 179, is more abundantly discovered than the second form in the same division of plants, and is believed by Prof. B. to be composed of crystals of the two first forms.

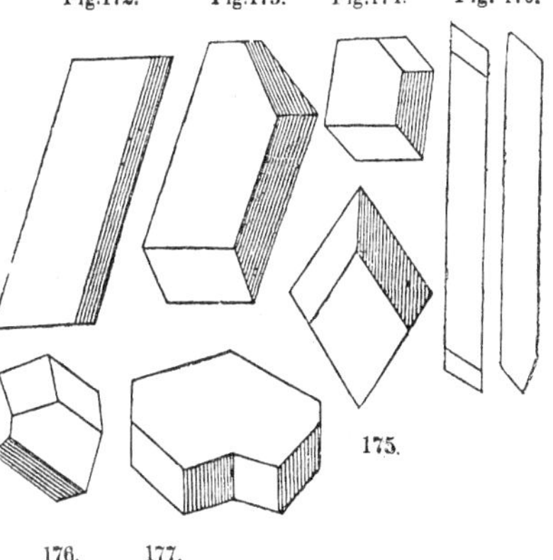
Fig.172. Fig.173. Fig.174. Fig. 170. 175. 176. 177.

The size of these crystals is very small, not being greater in some trees, as the locust, willow, &c., than the *twelve hundred and fiftieth of an inch* in length; but their number is so great that within the compass of a square inch of bark, not thicker than a sheet of writing paper, more than a *million* of these beautiful gems are collected together. And when we reflect, says Prof. B., "upon the number of such layers contained in the thickness of the bark, and the number of square inches given by the surface of a large tree, including all its branches, and then consider, that in addition to all this, the amount of crystals contained in the leaves, wood and roots is to be taken into account, we find that the number of crystals in a single tree is enormous beyond all conception. Yet the greater number of trees in the forests, not only in this but in all countries, are as full of these bodies as the specimens exhibited in figure 171." When the crystals found in wood are subjected to chemical tests they are generally found to be composed of *oxalate of lime.* Between the figures 178 and 179 a small scale of measurement is seen, which is magnified to the same extent as the crystals above described. The true length of the scale in the cut is *one-five hundredth* of an inch, and each division is equal in extent to *one-twenty five hundredth of an inch;* by comparing these divisions of the scale with the size of the crystals, their exceeding minuteness is at once recognised.

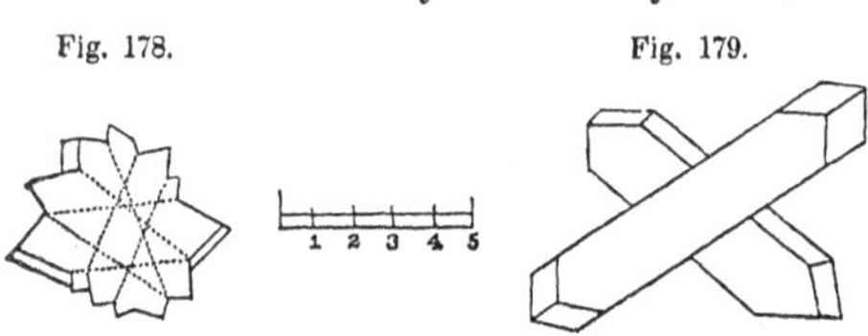

Fig. 178. Fig. 179.

In figure 180 is exhibited a collection of crystals obtained in the manner described by Prof. Bailey. Fixed in the indurated balsam they appear, in vast numbers, under the microscope, of a brown color, and bearing a resemblance in their shape to kernels of rice. A single crystal measures in length *one-six hundred and twenty-fifth part of an inch.*

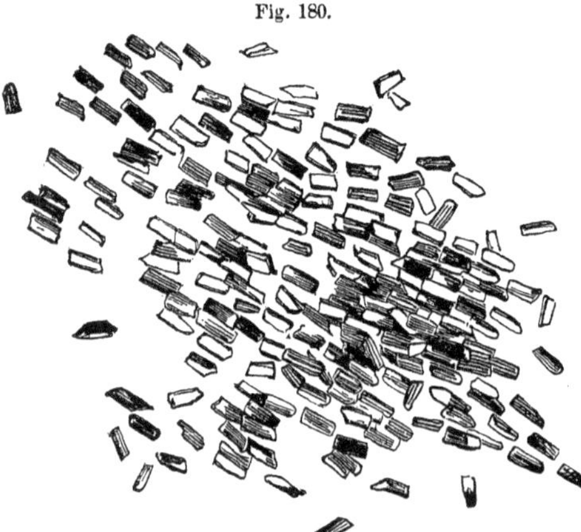
Fig. 180.

The delicate crystalline tracery existing in the burnt ashes of a maple leaf, is beautifully displayed in figure 181. The leaf from which the drawing was taken was prepared in the way above stated, when upon placing a portion of the ashes beneath the microscope, fine lines of brilliant crystals were beheld, as exhibited in the figure, following all the minute ramifications of the leaf. Some of the lines were more or less broken, on account of the fragile nature of the ashy film ; but many were preserved entire, and exhibited in the field of view, the crystalline network of the figure. Within the compartments formed by the chains of crystals, dark masses are beheld, which are crystals of a larger size. The small crystals measure in length *one-two thousandth part of an inch*, and the larger *one-seven hundred and seventy-fifth part of an inch.*

Fig. 181.

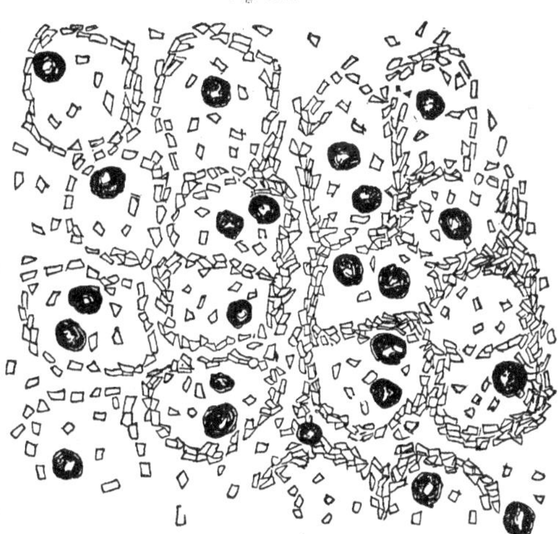

Amid the starch globules of potatoes and in the outer coating of the bulb of the onion, crystals have likewise been found, differing in form from all the preceding. In figure 182 the thin coating of an onion is delineated as it appears when magnified. This tissue is divided up into cells in which the crystals are formed. Their shape is that of a right square prism, and, in length, they measure *one-eight hundred and thirtieth part of an inch.*

Fig. 182.

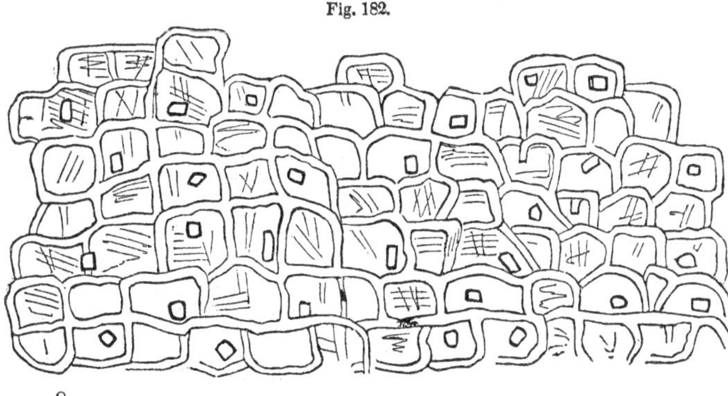

CHAPTER VI

PARTS OF INSECTS, AND MISCELLANEOUS OBJECTS.

Insects and mites, of mean degree,
That swarm in myriads o'er the land,
Moulded by Wisdom's artful hand,
And curled and painted with a various dye;
In your innumerable forms
Praise Him that wears th' ethereal crown,
And bends His lofty counsels down
To despicable worms.—WATTS.

EYES.—Nothing within the whole range of his investigations has more elicited the admiration of the philosopher, than the wondrous structure of the human eye. Exceedingly complex in all its arrangements, it abounds with exquisite contrivances for securing, under every circumstance, distinct vision; and so complete are the several parts in themselves, and so admirably adapted to each other, that it is justly deemed the most perfect of all optical instruments. Upon its curved and crystal front, fall the rays of light from unnumbered objects, spread over a landscape miles and leagues in extent; and the luminous lines converging in the eye with unerring accuracy to the interior surface, form a faithful picture of the entire scene, within the compass of a finger-nail. Perhaps a vast city is immediately before it, with its splendid panorama of towers and turrets, spires and cupolas, piles of massive buildings and thronged streets; while beyond, the harbor is crowded with the barks of commerce, and bays, and misty isles stretch away in the dim distance; yet all these are perfectly delineated upon the retina, in their just proportions and natural colors.

But if our wonder is excited when contemplating the structure of the eye of man, and of other animals, it is still more heightened upon examining the visual organs of insects, beneath the powerful glasses of the microscope. The eyes of insects differ from those of other animated existences, chiefly in respect to *number*, *form* and *arrangement*. In some, as in the spider, the number varies from six to eight, possessing such a diversity in their mutual arrangement, that their relative positions have been employed by writers to designate the several species. Thus, in one kind the eyes are arranged as in figure 183; in another as shown in figure 184; and in a third according to figure 185, and so on.

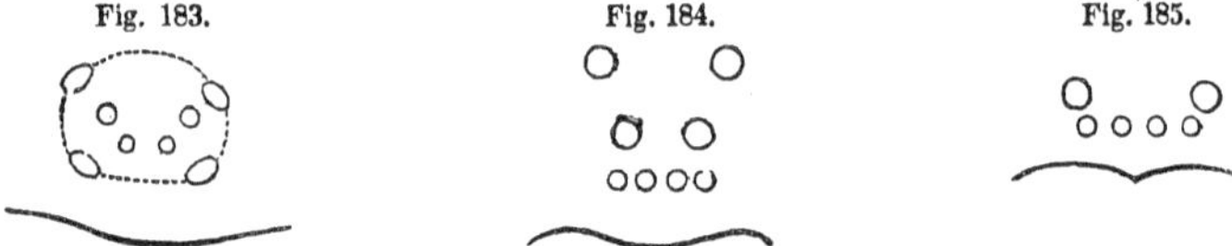

Fig. 183. Fig. 184. Fig. 185.

The scorpion has six visual organs, and the centipede twenty; but other insects, as the butterfly and dragon-fly, are gifted with a vast number of eyes, set in a common ball, to which the name has been given of *reticulated*, or network eyes. These complex organs appear to be designed for horizontal and downward vision; while *coronet* eyes are found placed upon the front and top of the heads of insects. These latter organs appear as round, transparent, and shining points, and are supposed to be employed for upward vision; they are usually three in number, and are generally arranged in the form of a triangle.

RETICULATED EYES.—When the eye of a butterfly or dragon-fly is viewed through a powerful microscope, it resembles a piece of network, and presents the appearance of a honeycomb; each apparent cell being a perfect eye. The outer surface of each is bright, polished, and round, like that of the human eye, and reflects as a mirror the images of surrounding objects. What therefore is commonly termed the eye of the dragon-fly, silk-worm, bee, and of other insects having similar organs of sight, is in fact a complex instrument of vision, consisting of a great number of single eyes, arranged in a globular case, each capable of forming distinct images of the objects before it. Dr. Hooke discovered no less than 7000 single eyes in the compound eye of a horse-fly, while according to the computation of Leuwenhoeck, more than 12,000 are contained in that of the dragon-fly; and M. Puget counted in each of the reticulated organs of some butterflies which he examined, the astonishing number of 17,325 lenses, each constituting a perfect eye. Optical artists have constructed an instrument called a *multiplying glass*, by taking a solid piece of glass, bounded on one side by a plane, and on the other by a curved surface, and then grinding and polishing the latter into a number of flat faces, still preserving, however, the general curvature. When a single object, as a flower, is beheld through this instrument, its images are multiplied in proportion to the number of exposed faces, and are all symmetrically arranged together, if the faces of the glass have been cut with regularity.

Reticulated eyes operate in the same manner · and naturalists, by carefully preparing these organs, and observing objects through them with the aid of a microscope, have been surprised and delighted at the wonders that have met their view. Not only are objects *multiplied*, but they are also *diminished* to a surprising degree. As Puget gazed at a soldier through the eye of a flea, an army of pigmies suddenly appeared before him, and the flame of a candle flashed forth with the splendor of a thousand lamps. When Leuwenhoeck, in like manner, directed his sight to the steeple of a church two hundred and ninety-nine feet high, and distant seven hundred and fifty feet from the place where he stood, it appeared no larger than the point of a cambric needle.

The reticulated eyes of many flies shine with the brilliancy of the finest gems, and gleam with the richest hues of light. In some the tints are red, in others green, while a third class glow with a play of colors of surpassing beauty, formed of mingled yellow, green and purple. Some ephemeral insects are gifted with no less than four of these wonderfully complex organs, the ordinary pair being of a brown color, while the additional pair, shining with a beautiful citron hue, rise side by side from the upper part of the head. The form of the single lenses in reticulated eyes is not the same in every insect endowed with this curious organ; for in the compound eye of the dragon-fly and honey-bee, the lenses are six-sided; while in that of the lobster they possess a square form. In figure 186 is shown a portion of the cornea of the compound eye of a dragon-fly, the single eyes of which are seen to be six-sided, and regular hexagons. In certain positions, in respect to the direction of the light, they gleam with a rich golden hue, and three parallel borders are discerned, which divide the single eyes from each other. The inner circle in figure 187 represents the same object of its natural size. Figure 188 presents a magnified view of a part of the complex eye of a lobster, composed of a great number of single eyes, possessing a square form; the real size of the object is shown by the smaller circle in figure 189.

Fig. 186.

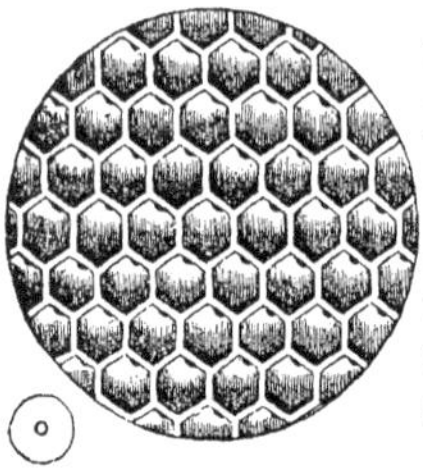

Fig. 187.

Fig. 188.

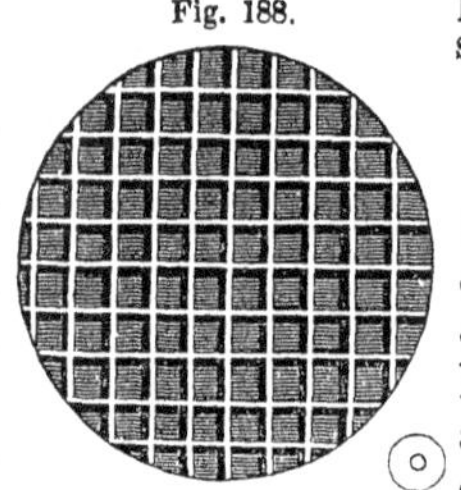

Fig. 189.

The eyes of the bee, which are delineated in figure 190, are described by Swammerdam as being profusely covered with hairs, which pierce through the outer covering of the eye, in the same manner as the hairs of the human body penetrate through the skin. These hairs are very numerous, bristling in thick profusion over the eye, and are supposed to perform the office of eye-lashes or eye-brows, in protecting the organ from dust, or any similar annoyances that might work it harm. In this figure, the compound eyes of the bee, with the parts adjacent, are beautifully and distinctly revealed.

The upper part of the wood cut exhibits one of the eyes in its perfect state, composed of hexagonal lenses, and bristling with hair. In the lower portion of the same figure, the other complex eye is shown, deprived of some of its hexagonal lenses in order that its structure may be perceived: the lenses or single eyes are here seen to have the shape of a pyramid. The three oval figures, situated together in the angle formed by the two compound eyes are the *coronet eyes* of the insect, while the two branching members that curve over the reticulated eyes, are the antennæ of the bee. Between these the head is thickly covered with plumes of hair. Figures 191 and 192, represent seven of the hexagonal lenses, very highly magnified, and which, in 192 are exhibited bristling with hair.

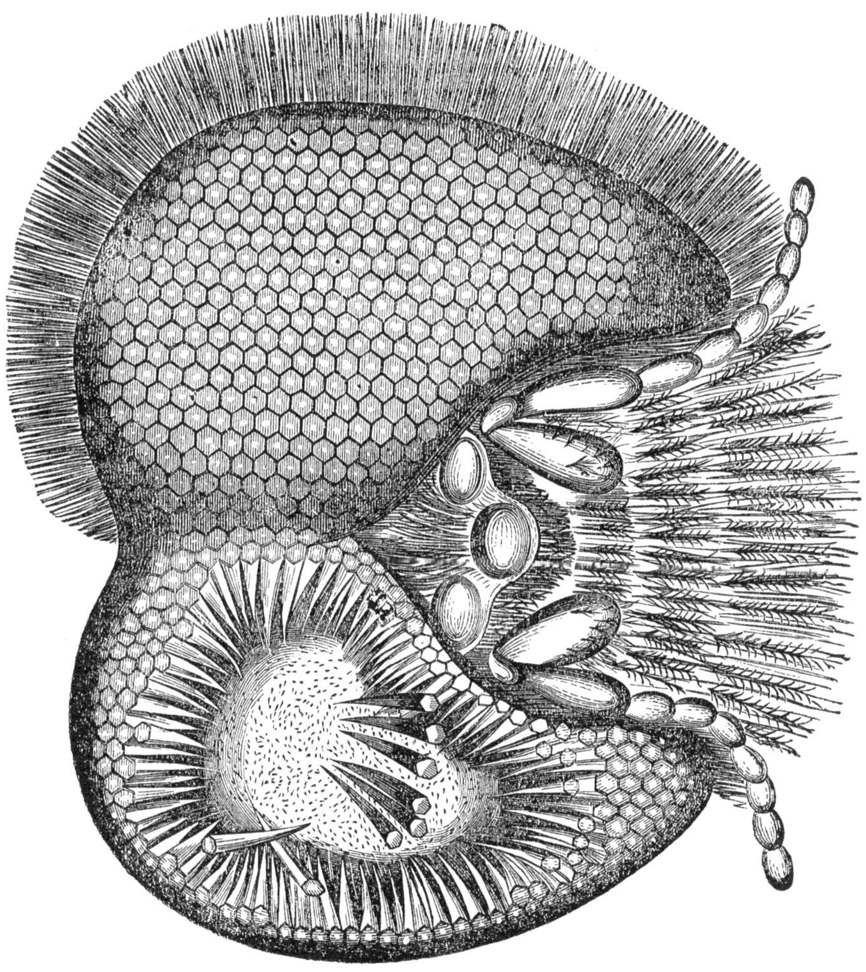

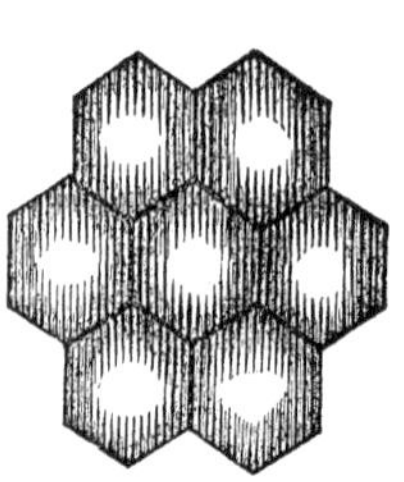

Fig. 191.

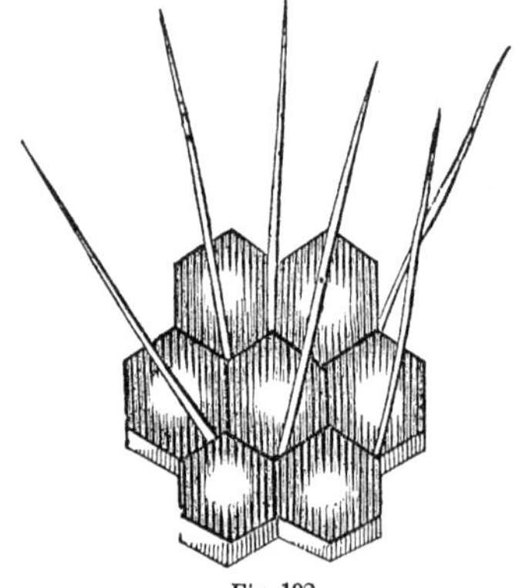

Fig. 192.

WINGS.—The wings of insects afford curious and interesting objects for microscopical examination; since in form and structure their diversity is endless, and their rich adornments and exquisite hues are often surpassingly beautiful. When magnified, a great number of minute joints are brought to view, by means of which the gauze-like wings of many of these little creatures are at one time curiously folded up within their shelly cases, and at another are instantly expanded for flight; while numerous branches of veins, nerves and muscles extend throughout these delicate structures, conveying life, strength, and action to every part. The hard and shell-like cases, under which these transparent wings are securely folded, are usually highly polished, and are often adorned with elegant flutings, and a rich diversity of splendid tints. The diamond-beetle possesses all these beauties, and is regarded as one of the most brilliant objects in nature. Its head, wings, and legs are studded with scales, glowing with the resplendent colors of the sapphire, ruby, and emerald; and it is said, that in Brazil, where they are found, such is the dazzling splendor of their hues, that the eye cannot endure their radiance as they fly in swarms through the air upon a sunny day. In figure 193 the wing of an earwig is shown, of its natural size, and the same is also there delineated as it appears when magnified. The upper part of the large cut represents the wing-case, which is opaque and shelly; while the rest of the figure exhibits the greater part of the wing, which is thin and transparent, and folds up neatly beneath the wing-case, which is not more than one-sixth of the entire wing in size. Some of the ribs are seen radiating to the border, like the sticks of a fan, from a small space in the upper part of the wing, while others intervene of shorter length, and proceed from the margin half way towards this central spot. All these ribs are connected together by a band that runs along parallel to the margin; the entire arrangement being evidently contrived so as to impart, at once, strength and lightness to the wing, and thus facilitate its rapid motion. When the wing closes, the insect first turns back the marginal part, and then closes the ribs in the manner of a fan, folding up, within a small compass, the entire delicate structure of the wing, under the protection of the strong shield of the wing-cover.

Fig. 193.

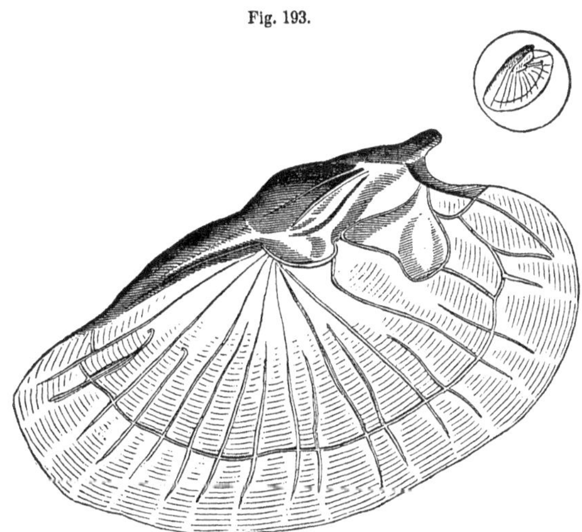

The wing-cases are not in all instances composed of a horny substance; since among the beetle tribe they frequently consist of a softer material like leather.

When the insect is preparing to fly, the wing-cases are opened to such an extent as to allow full play to the wings; the insect then launches into the air, striking it vertically with these delicate organs, while the wing-cases are kept immoveable during the whole time of flight. The resistance presented to the atmosphere by the latter is supposed to facilitate in some way the motions of these little beings. The bodies of insects, like the beetle, are almost in an upright position, during their flight, and present a singular appearance, in the case of the larger kinds, as they move heavily and laboriously along. The wings of the beetle are for the most part of great extent, and the ribs that ramify all over their surface are stronger than those which are found in the wings of other orders of insects; and are so arranged as to strengthen and support every part. In addition to what has already been remarked regarding the structure of the wings of insects, it may be further observed, that the ribs are hollow tubes originating in the trunk, and that within them are tubular vessels, which are supposed to be air-vessels communicating with the organs of respiration in the trunk.

HEMEROBIUS PERLA.*—In cut 194 is delineated the wing of the Hemerobius Perla, both magnified and of its natural size. This insect receives its appellation from the short duration of its life, which lasts but two or three days. Its wing is extremely elegant and delicate, and is formed of a membrane as thin as the finest gauze; while slender ribs, fringed with hairs of a greenish tinge, are seen strengthening the wing and running both lengthwise and obliquely to the margin. From these ribs lateral branches proceed, in directions for the most part, parallel to each other; and from the latter a third series arises; the whole forming a strong and compact network bound firmly together. In addition to the beauty and regular structure of its wings, this insect is otherwise adorned, its body being tinged with a delicate green, while its two eyes glitter like beads of polished gold. A striking instance is here afforded of the care bestowed by our Heavenly Father upon one of the smallest and least enduring of his works; for the Hemerobius in the course of a few hours comes into being, matures and dies; completing, in this short space of time, the whole round of its existence, and yet its Creator has not only bestowed upon it those organs and powers which are necessary for the discharge of the various functions of its life, but has not deemed it beneath him to lavish upon it the bright gifts of beauty.

Fig 194

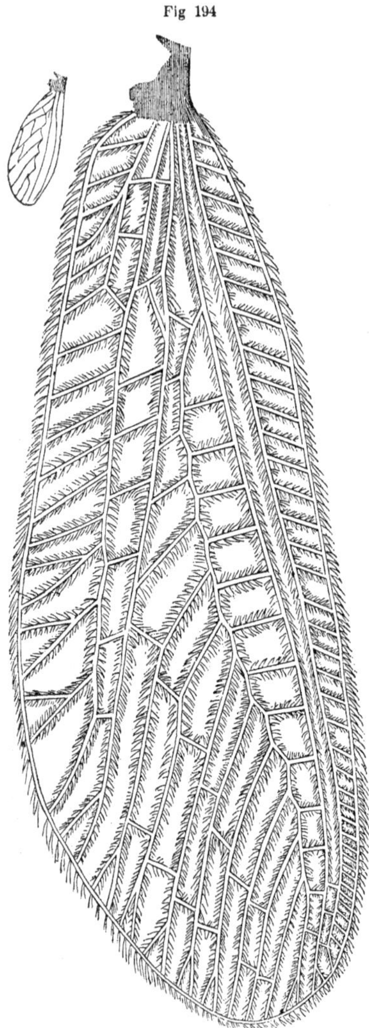

FEATHERS OF MOTHS AND BUTTERFLIES.—A certain order of insects, which includes Moths and Butterflies, have received the name of Lepidoptera† from the peculiar construction of their wings. These members are covered with a *fine*

* From the Greek hemera, a *day*, and bios, *life*.
† From the Greek *lepis*, a scale, and *ptera*, wings.

PLATE IX.

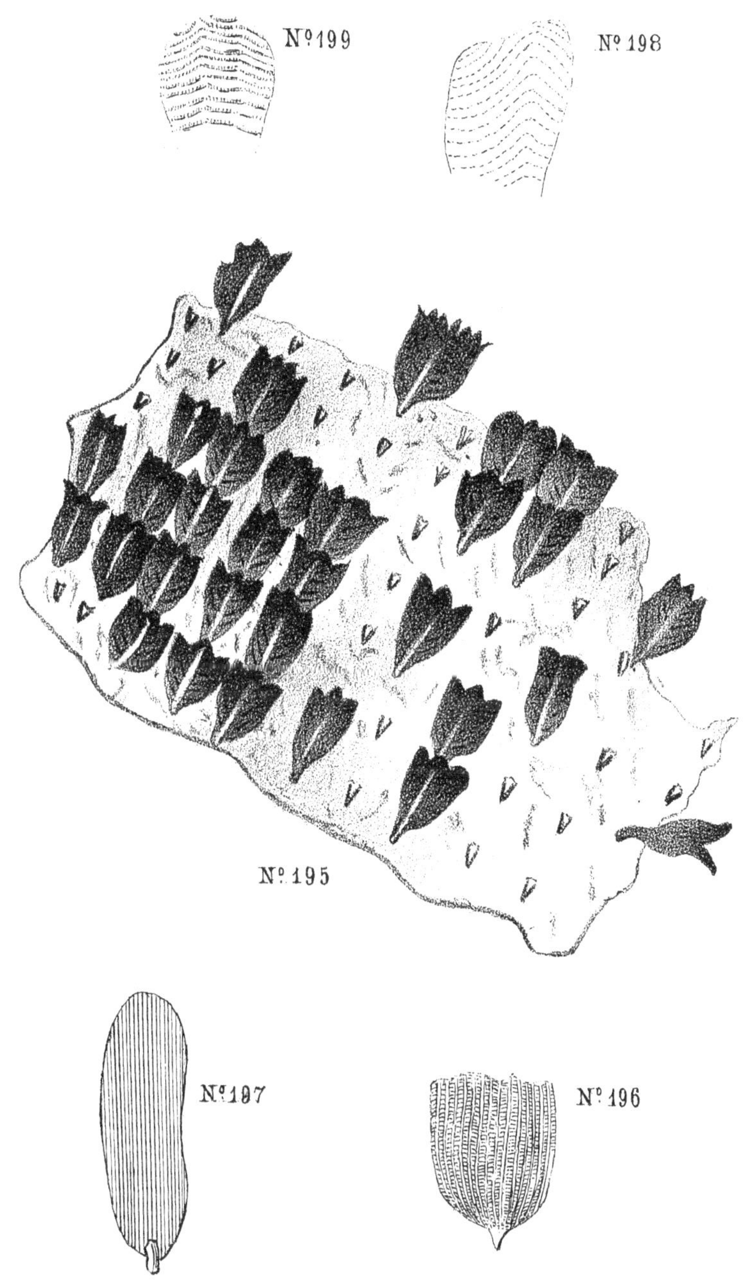

dust, resplendent with the most brilliant colors, adorning their variegated surface. When this dust is examined by the microscope, it is discovered to consist of a vast number of minute *feathers or scales*, differing in form, and as remarkable for the elegance and symmetry of their structure as for the beauty of their hues. Their shape not only varies in different insects, but the same insect possesses feathers of different forms. Some are long and slender, others short and broad; these are smooth at the edges, and those serrated or notched; one kind is triangular and a second oval. When viewed under the microscope, these scales are found to be terminated by a short stem, that connects them with the wing, and their surfaces are grooved in lines and stripes, which take different directions in different scales. In some feathers two or more sets of lines are discerned crossing each other. When this feathery dust is brushed away from the wing of a butterfly, the surface below appears like the wing of the Hemerobius—a network of ribs connected by a delicate, transparent, and elastic membrane. The ribs are hollow, by which contrivance, a wing, though broad and extended, still retains its lightness, and on the membrane rows of dots are perceived, where the stems of the scales were attached to its surface. On those parts where the dust remains, the little particles are seen magnified into feathers, symmetrically arranged and overlapping each other like the scales on a fish. These several appearances are beheld in drawing 195, which represents a portion of the wing of a butterfly, called the Papilio Archippus, partially divested of its scales. In this magnified view the lighter parts of the drawing represent the delicate exposed membrane of the wing, while in the rest of the figure it is covered with scales, which are of a rich brown hue, partially overlapping each other in regular rows. The several lines of dark spots that cross the membrane of the wing are the places where the stalks of the feather were fastened to its surface. The breadth of these spaces is the *sixteen hundred and sixty-sixth part of an inch*, and the width of one of the scales *the two hundred and fiftieth part of an inch.* The scales on the wings of an insect, termed the Lepisma Saccharina, are of two kinds, one set being arranged as usual in rows, and the other, possessing a different shape, are inserted *between* and *over* the former, fastening them firmly down in their places. In some instances the scales are distributed over the membrane without apparently any regular arrangement. Drawings 196 and 197 exhibit magnified views of scales taken from the wing of a butterfly, known by the name of the Morpho Menelaus. The color of the upper surface of its wing is of a rich blue, brilliant beyond description, and vying in splendor with the purest azure of the sky. The scales taken from the central portions are of a pale blue tint, mingled with others that are almost black. The former are for the most part wider than the latter, and measure nearly the *one hundred and twentieth part of an inch in length.* Beneath the microscope they exhibit the appearance presented in drawing 196; the entire surface being fluted with lines which run lengthwise of the scale, and are connected together by short cross lines passing between them. In drawing 198 and 199 are delineated the feathers of the *lead-colored* Spring-tail, an active little insect about the tenth of an inch long, found among saw-dust and damp wood. Its body and limbs are cased in delicate scales, varying from *one-nine hundredth to one-one hundred and sixtieth part of an inch in length;* and which are covered with lines diversified in arrangement as displayed in the above-named figure. Scales of very singular form are found on the under side of the wing of a beautiful blue butterfly, called the Lycœna Argus. In shape they resemble a battledore, with strings of beads running lengthwise over the surface parallel to each other. The feathers upon the wing of an insect are exceedingly numerous, for according to Leuwenhoeck each wing of the moth of the silkworm contains more than *two hundred thousand,* and the wing of this insect is small compared with that of many other moths; for one of the largest, the Atlas moth, the feathers of which are delineated in figure 200, measures nearly a foot across the wings. In figure 201 is delineated

Fig. 200. Fig. 201.

several scales from the wing of the Death's-head moth, which receives its name from bearing upon the surface of its thorax a large grey or yellowish spot, which strongly resembles in form the front view of a human skull or Death's head. On account of this peculiarity, and also from its great size, and the power it likewise possesses of emitting a plaintive cry, this insect has been regarded with superstitious awe. Reaumur relates, that appearing once in great numbers in some districts of Bretagne, they were viewed with terror by the inhabitants, as the sure precursor, and even the cause, of war and pestilence. In German Poland it is termed the Wandering Death Bird; and to the dreamy imaginations of the superstitious, the head of a perfect skeleton is distinctly visible on the insect, resting upon the limb bones crossed beneath; while its cry is the moaning of a child in pain and suffering. Its creation is regarded as the work of evil spirits, and the reflections of those lurid flames amid which it arose, are discerned in the gleaming of its glittering eyes. The rich hues of the scales of the Lepidoptera are supposed in general to be due to the presence of coloring matter, but the more delicate tints are regarded by Dr. Roget as an optical effect, produced by the fine lines upon the surface of the scale; a phenomenon identical with that observed in mother of pearl, where the concentric flutings of the shell occasion the brilliant play of colors that adorn its surface.

EGGS.—The eggs of birds, though differing in color, possess nearly the same form, varying only by slight changes between an oval and a globular shape. Such however is not the case with those of insects, which exhibit an endless variety of exquisite forms often most beautifully and elaborately wrought, and bearing a resemblance to richly carved work. "We meet with them," says Kirby, "of the shape of the common hen's egg, flat, round, elliptical, conical, cylindrical, hemispherical, pyramidal, square, lens-shaped, turban-shaped, pear-shaped, boot-shaped, and sometimes of shapes so strange and peculiar, that we can scarcely credit their claim to the name of eggs." Indeed, the empty shells left upon the leaves of plants have been mistaken for minute flower cups, and according to Reaumur were once actually thus delineated by a naturalist, who was extremely perplexed to account for the origin of these singular blossoms. Among the most rich and elegant forms, are those which belong to the eggs of butterflies, the surface being often exquisitely sculptured and profusely adorned with ornaments. Four varieties are delineated in drawings 202 and 203 as they appear when highly magnified.

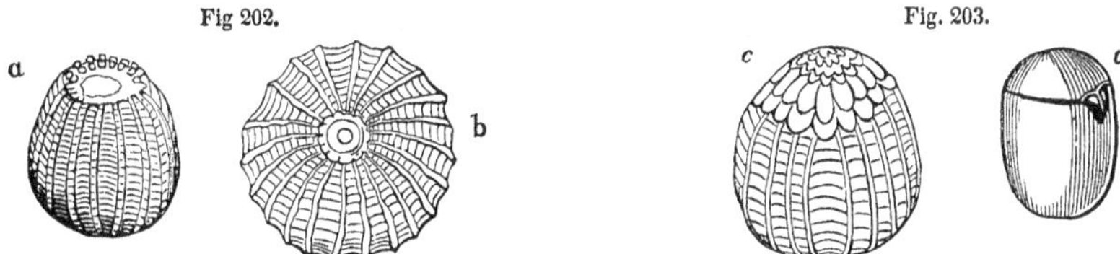

Fig 202. Fig. 203.

In this group, *a* 202, is the egg of a butterfly called the Hipparchia Tithonus; it is of a dome-shaped form, strengthened and adorned with longitudinal ribs, symmetrically arranged and connected by cross lines; *c* 203, is the egg of another kind, the Hipparchia Furtina, and is crowned at the top with circular rows of scales, overlapping each other likes the plates of armor or the scales of fishes. The same type is observed in the egg of the Noctua Nupta, at figure *b* 202, where the end of the egg is presented to view; and towards which numerous strongly defined ridges converge. Between these ridges the surface is deeply fluted. The eggs of many insects are provided with a small lid or cover, which when the young insect within has arrived at maturity it throws off at its will, and emerging, through the opening thus made, from the enclosing shell, enters at once upon its new state of existence. In addition to this provision, a curious contrivance is found in the egg of a certain species of bug, and which is shown at *d* 203. It consists of a horny substance in the shape of a cross-bow, the bow being attached to the lid, and the handle to the upper end of the side of the egg. It is supposed to be designed to facilitate the egress of the young when it is ready to leave the shell.

HAIRS.—The hairs of different animals afford beautiful objects for microscopical examination. Those of the common mouse which are shown in drawings 204 and 205, vary in form and size, their diameters ranging from *one-two thousandth* to *one-three hundredths of an inch.* The chief varieties are here exhibited of their relative sizes in figures 204 *a* and *b*, and 205 *a*, and are delineated as they are seen by transmitted light; the real diameter of the hair in cut 204 *a*, is only *the sixteen hundredth part of an inch.* When the hairs are viewed by reflected light, their appearance is changed, for the solid parts then reflect more light than the transparent, and appear white, while the transparent portions are comparatively dark. The appearance presented by a magnified hair when viewed by reflected light is shown in figure 205 *b*. The hair of the bat is different from that of the mouse, and consists of many varieties distinguished from each other in form and structure. The two principal kinds are delineated in figures 206 and 207. The first represents a collection

Fig. 204. Fig. 205.

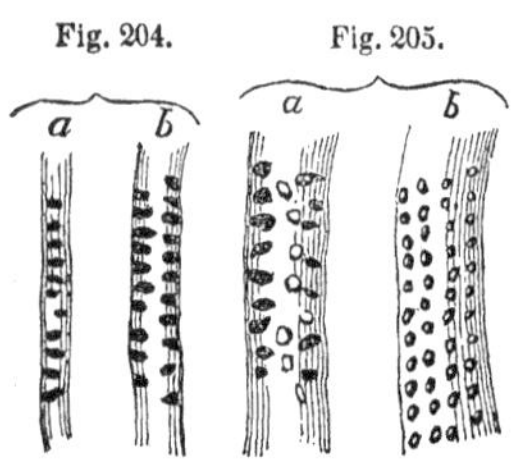

Fig. 206. 207. 209. 208.

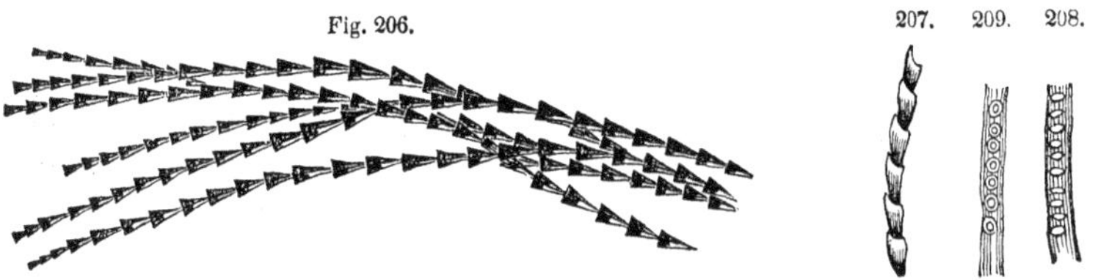

of hairs scattered promiscuously together, each possessing a figure like that which would be formed by a series of cones with the points of each inserted into the middle of the base of another. The second exhibits a curious spiral structure. Figure 208 is a white hair from a young cat; figure 209 that of a Siberian fox, and 210, the hair of a common caterpillar, which divides into

Fig. 210.

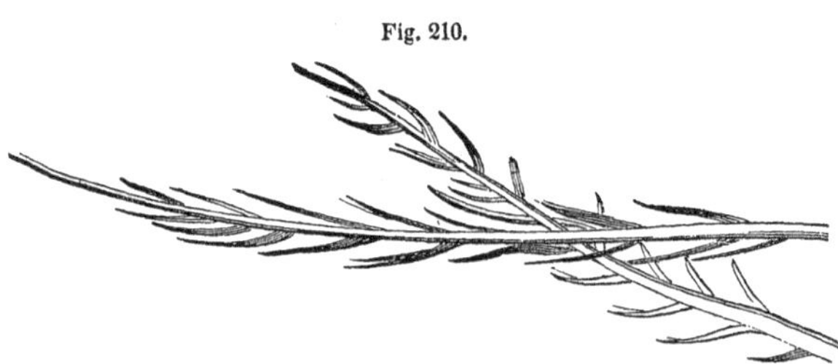

branches; but the form of these hairs is different for every species of caterpillar; in some they resemble the spreading plumes of the peacock's tail, while others are adorned with delicate tufts of hair, and bristle with thorns.

THE PROBOSCIS OF THE OX-FLY.—This insect, which is the torment of cattle during the summer months, and nourishes itself upon their blood, is provided with a curious apparatus admirably adapted for piercing the tough hide and imbibing the blood of its victim. The proboscis is shown, of its natural size, in figure 211; and a magnified view of the same is presented in figure 212.

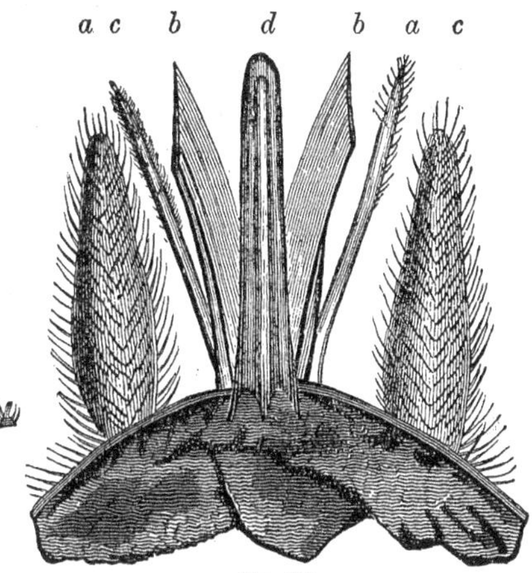

Fig. 211.

Fig. 212.

This member is complex in its structure, and is enclosed in a fleshy case, which is removed in the drawing, in order that the several parts may be separately exhibited. The parts are *six* in number, besides the two feelers, *a*, *a*, which are composed of a spongy substance, fringed with hair. They are of a gray color, and are capable of motion, each being furnished with a joint, at the point where it is connected with the head.

The office of these appendages is to protect from harm the delicate parts of the proboscis, since they are always placed on each side of it whenever a puncture is made by the insect. The blades or lancets, *b*, *b*, are the instruments which inflict the wound: in shape they are alike, each having the form of a broad knife with a sharp point and fine edge, and gradually increasing in thickness towards the back. The parts, *c*, *c*, termed *piercers*, are furnished with teeth like a saw, and are supposed to be employed for the purpose of increasing the size of the wound, and thus obtaining a more copious supply of blood. It is also imagined, from being of a hard texture, that they likewise serve as a protection to the tube which conveys the blood from the wound to the stomach of the insect, and which would otherwise be liable to receive injury. This tube is seen inclosed in its fleshy case, *d*, with a lancet and serrated piercer upon either side.

THE SUCKER OF THE GNAT.—The sucker of the gnat appears to the naked eye like a sharp needle, finer than a hair, but under the microscope it presents a complicated structure; and although it has been minutely examined, the most distinguished observers have differed in respect to the number of parts of which it is composed. Leuwenhoeck enumerates *four* parts, Swammerdam *six*, including the lip, and Reaumur *five;* but it has been supposed that their observations might possibly have been made either upon mutilated insects, or upon those of different species. As soon as a gnat has settled upon some exposed part of an animal, it puts forth from the sheath of its sucker a fine point, with which it pierces the skin. This point was regarded by Swammerdam as single, inasmuch as he was unable to discern the least breadth at the extremity, under the best microscopes at his command, but both Leuwenhoeck and Reaumur discovered it to consist of several needles, some of them barbed and serrated.

In fact, the compound structure is revealed upon pressing the piercer between the finger, when the several parts separate from each other. These fine needles are inclosed in a sheath formed of some yielding substance, and divided throughout its whole length; and not only does it shield these slender instruments from injury, but also serves to support and steady them during the operation of penetrating the skin; answering the same purpose as the fleshy protuberances in the proboscis of the ox-fly. The sheath is also supposed by Swammerdam to be employed as a tube, through which the insect imbibes the blood that flows from the wound.

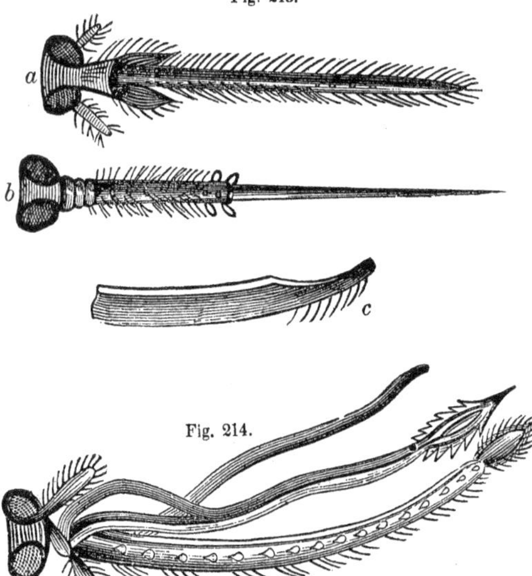

A magnified view of the several parts is exhibited in cut 213, where *a* represents the sucker in its sheath; *b*, half of the sheath broken off, in order to show the sucker; and *c*, the barbed point of one blade of the sucker. In cut 214, the sucker is displayed so as to exhibit its component parts.

The Proboscis of the Bee.—The exquisite instrument, by which the bee collects from the flowery realm the rich nectareous food that is necessary for its support, is a most elaborate structure, and every part admirably subserves the purposes for which it was made. It has been most carefully analyzed both by Reaumur and Swammerdam; and the latter observer was so struck with the proofs of wisdom and benevolent design, revealed in this minute member, that at the close of his investigation he breaks forth into the following pious strain: "I cannot refrain from confessing, to the glory of the Incomprehensible Architect, that I have but imperfectly described and represented this small organ; for to represent it to the life in its full perfection, as truly most perfect it is, far exceeds the utmost effort of human knowledge."

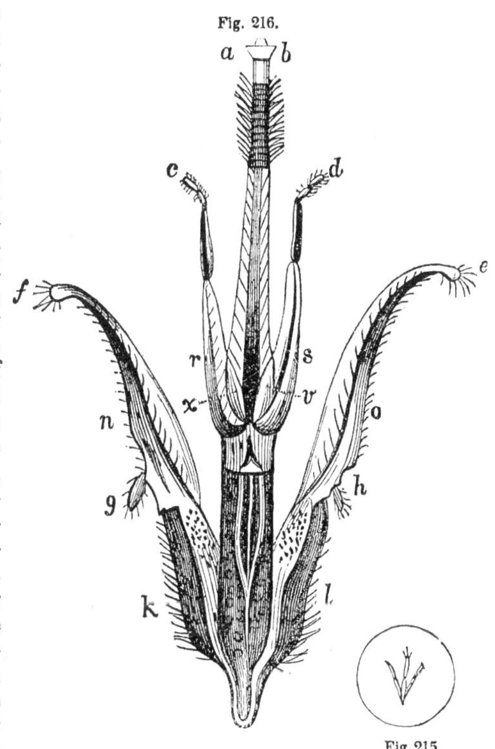

Fig. 216.

Fig. 215.

The proboscis of the bee is shown of its natural size in figure 215, and a magnified view of the same is presented in figure 216. It is here seen to consist of five distinct parts, the central stem, *a*, *b*, called the *tongue*, and four other parts arranged in pairs, constituting two sheaths. The exterior sheath is formed of the two branches, *f*, *n*, *g*, *k*, and *e*, *o*, *h*, *l* ; and the interior of the parts *c*, *r*, *x*, and *d*, *s*, *v*. These sheaths are composed of a horny substance, are fringed with hair, and provided with joints; and fold down upon the tongue one over the other, forming together in appearance a single tube, convex outward, and concave inward, towards the trunk of the bee. The articulations of the outer sheath are at *g* and *h*, and the parts above, which in the figure are widely separated, can be folded down at pleasure upon the central stem, by means of these joints. The branches of the interior sheath are each possessed of three joints, the lower jointed portion being longer than either of the other two, which are always kept curved outwards, as represented in the figure at *d* and *c*, even when the complex apparatus is closed as much as possible. It is supposed by Swammerdam, that these fringed joints aid the bee in the manner of fingers by opening those flower-cups that but partially reveal their sweets, and removing obstructions that would otherwise prevent the tongue from reaching the inmost recesses of the blossom. The upper part of the tongue consists of rings, fringed with circles of hair, and it terminates in a small knob, which is also fringed. The office of the hair is to brush off and secure the honey discovered by the insect in the flower-cells. The lower part of the tongue is membranous in its structure, and can be greatly distended, and when the bee is collecting its food, form a capacious bag for the sweet juices, that are ultimately converted into honey. The insect gathers its treasures by lapping them up with this complex instrument from the nectaries of flowers, sweeping with equal facility the round or the concave surface of the leaves. When a sufficient portion is thus collected, it is first deposited in the reservoir just mentioned, and thence conveyed to the honey-stomach. As soon as the bee has rifled a blossom of its honey, the several branches of its proboscis are quickly folded together to protect the more delicate parts from injury, and when it is again to be employed, they are as rapidly again expanded. While at rest it is doubled up by means of a joint, one branch being brought within the lip, and the second secured beneath the head and neck.

Sting of the Wild Bee.—The sting of the wild bee, with its several parts, is delineated in figure 217, copied from an engraving of the drawings of Mr. Newport, who examined and dissected this organ with the utmost care and describes it as follows: "The sting is formed of two portions placed laterally together, but capable of being separated: *b* is the sting, the point of which is bent a little upward, and becomes curved, as shown at *d*, where the barbs are exhibited more highly magnified. They are about six in number, and are placed on the under surface, with their points directed backwards. At the base of the sting, *e*, there is a semi-circular projection, apparently for the purpose of preventing the instrument from being thrust too far out of the sheath in which it moves; it has likewise a long tendon to which the muscles are attached. Between these parts, (the sting and the sheath,) when brought near to each other, the venom flows from the orifice at the extremity of the poison-tube, which comes from the anterior portion of the poison-bag, *a*. This bag is of an oval shape, and is not the organ which secretes the venom, but is merely a receptacle for holding it, since it is conveyed into this reservoir by means of a long winding tube, which receives it from the secreting organs at *f*. The tubular sheath of the sting is seen at *c*; it is open at its base and along its upper surface, as far as the semi-circular projection before mentioned. The muscles which move the sheath are distinct from those which give motion to the sting. The sting of the wild bee resembles that of the honey-bee."

Fig. 217.

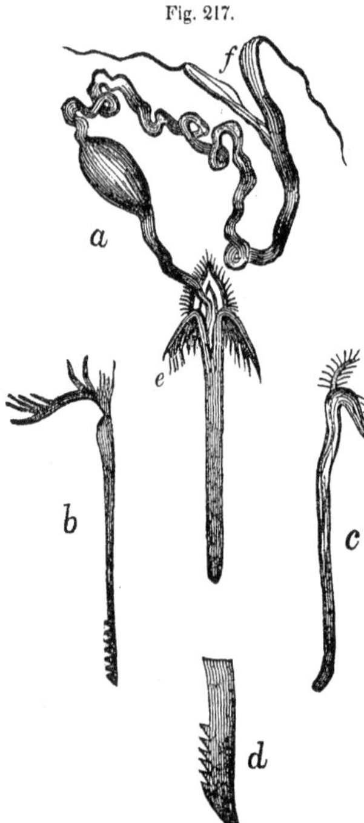

Feet.—The last joint of the feet of insects is usually terminated by a claw, either single or double, and in the case of spiders it is divided into three branches.

A triple claw of a spider is exhibited in figure 218; the several hooks are not smooth, but are armed with teeth on one side, and present quite a formidable appearance. The length of the tooth, *a*, the third from the end, on the middle hook, is the *five hundredth part of an inch.*

Fig. 218.

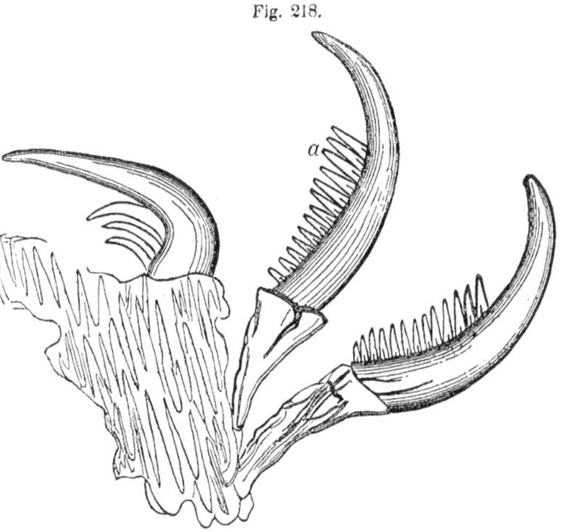

By the aid of their claws, insects are enabled to move over rough substances with great facility, either upward or downward, but upon polished surfaces they advance with the utmost difficulty. Upon the foremost pair of their feet these hooks are bent *backward*, on the posterior pair *forward*, and on the third or middle pair, *inward;* thus rendering the position of the insect exceedingly stable, and effectually securing it from displacement. The claws are, therefore, highly useful to the insect, either in a state of rest or activity. On the feet of the larger kinds of insects, cushions, composed of thick tufts of fine hair, are found, which prevent it from receiving injury upon leaping from a considerable height. Moreover, these delicate and elastic hairs adapting themselves to the asperities on the bodies which the insects frequent, enable the latter to adhere to them with much tenacity. But a more efficient apparatus is possessed by some of these little creatures, which gives them the power of walking in any position upon smooth and glassy surfaces. It consists of *suckers*, so adapted to the foot, that the insect is sustained by the pressure of the air upon the sucker.

The sucker consists of a thin membrane, capable of expansion and contraction, having the edges serrated or notched, so that it can be made to fit closely to surfaces of every shape. The sucker acts in precisely the same manner as the circle of leather with a string attached to the centre, which lads use in their sports to take up stones and pebbles, the leather being first wetted in order to make it adhere closely to the stone. The common house-fly has two suckers to each foot, immediately under the root of the claw, and attached by a narrow neck capable of motion in all directions. These appendages are delineated in figure 219, which represents the suckers on the under side of the foot of a blue-bottle fly, with the claws of the insect branching over them. In the horse-fly, every foot is provided with three suckers; and in the yellow saw-fly, four are arranged along the under surface of the toes, one upon each of the four first

Fig. 219. Fig. 220.

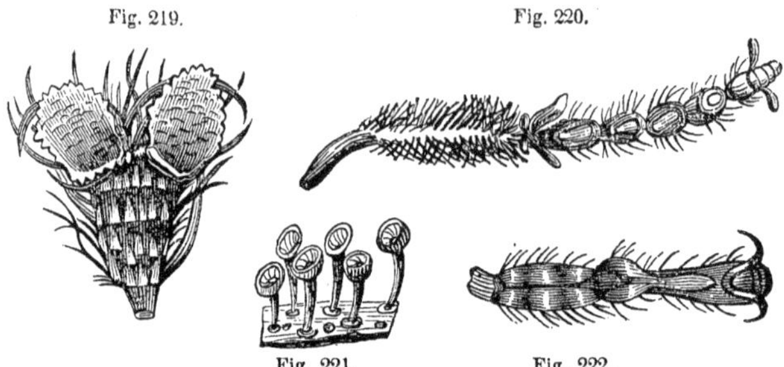

Fig. 221. Fig. 222.

joints, as in figure 220. In a species of water-beetle, the male insect is alone provided with a numerous collection of suckers. The first three joints of the feet of the fore-legs, have the form of a shield, the under surface of which is covered with suckers, some very large, others small, and a third class exceedingly minute, all provided with long, hollow stems. Several of the smallest kind are exhibited in figure 221, highly magnified, and having the appearance of mushrooms with the cups inverted. The corresponding joints of the second pair of feet, are likewise studded on the under side with a vast number of minute appendages of this character.

A certain species of grasshopper, called the Acridium biguttulum, is furnished with a large oval sucker, which is placed between the claws beneath the last joint of the foot. The first joint is padded on the lower side with three pairs, and the second with one pair of cushions. These cushions are filled with an elastic fibrous substance, the texture of which is looser towards the margin, in order to increase the elasticity. The several parts are displayed in figure 222. By the aid of this singular apparatus, insects are enabled to traverse with the greatest facility the smoothest surfaces, in an inclined, vertical, or inverted position. Thus a fly is seen to walk upon a mirror, a ceiling, or the under surface of a pane of glass, with as much ease and security as upon the top of a table. Indeed, when inverted, the weight of the fly causes the sucker to adhere more firmly than in any other position ; inasmuch as the weight of the insect tends to draw down the sucker and to increase the vacuum beneath it, and thus to render the pressure of the atmosphere upon the sucker proportionally greater. To this fact has been attributed the circumstance, that flies congregate upon the ceiling and repose there during the night, since the pressure of the air upon the membrane of the suckers fixes them firmly in their resting-place, without any voluntary effort.

Antennæ.—This name is given to certain curious organs possessed by insects and crustaceous animals ; the majority of the latter class being endowed with four, while no insect has more than one pair. They are inserted in the head, and, except where the insect has four eyes, are either placed in the space between the eyes, or in that immediately beneath them. They are, for the most part, formed of a number of tubular joints, so as to admit of motion in all necessary directions. The office of these organs is not yet thoroughly understood. Many celebrated naturalists consider them merely as feelers, that serve to guide the movements of the insect; while others, no less distinguished, consider this office as but secondary, and that the primary use of the antennæ is to enable the insect to detect *sounds*—that they are in truth *organs of hearing*. The forms of the antennæ are extremely various, and many of them are quite elegant and beautiful. In cut 223, a few only of these organs are delineated; one, that of

Fig. 223.

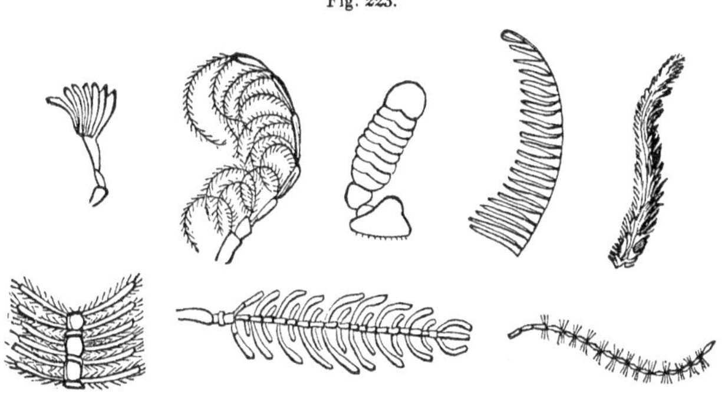

the common cockchaffer, is seen expanded like a hand with seven fingers; another exhibits the form of a graceful plume, and a third bears some resemblance to a feather. A fourth variety is thick and bushy, like the tail of a cat; some are fringed with slender arched branches, and others exhibit the form of a string of delicate beads studded with minute tufts of hair. Doubtless the great diversity, in form and structure, which obtains in these singular organs is needed for the well-being and enjoyment of the different little creatures to which they belong: each change in form, size, and figure, or in any other particular, being subservient to some wise and benevolent purpose; and were we but able to explore the whole field of research, we should be enabled to trace the hand of Divinity in every minute modification.

Scales of Fishes.—The scales of fishes furnish a great variety of beautiful objects for the microscope; their figures being often extremely elegant, and presenting a rich diversity of forms. For not only are different fishes possessed of different shaped scales, but those that belong to the same fish vary in structure, according as they are found on one or another part of the body. Leuwenhoeck supposed that each scale was composed of a vast number of minute scales, in rows, one layer overlapping another, the largest being next to the fish, and the rest gradually diminishing in size; thus forming successive strata from the base to the upper edge of the scale. In some scales, when viewed by the microscope, a great number of concentric flutings and grooves are discerned, too fine and too near each other to be distinctly counted, which are formed by the edges of the strata; each line denoting, as is supposed, the margin of each stratum and the different stages of growth in the scale. These flutings are often crossed by others proceeding from the central portion of the scale, and terminating at the circumference. The next twelve figures exhibit the structure of the scales of several fishes, most of them well known. In figure 224, is delineated the scale of a species of

Fig. 225.

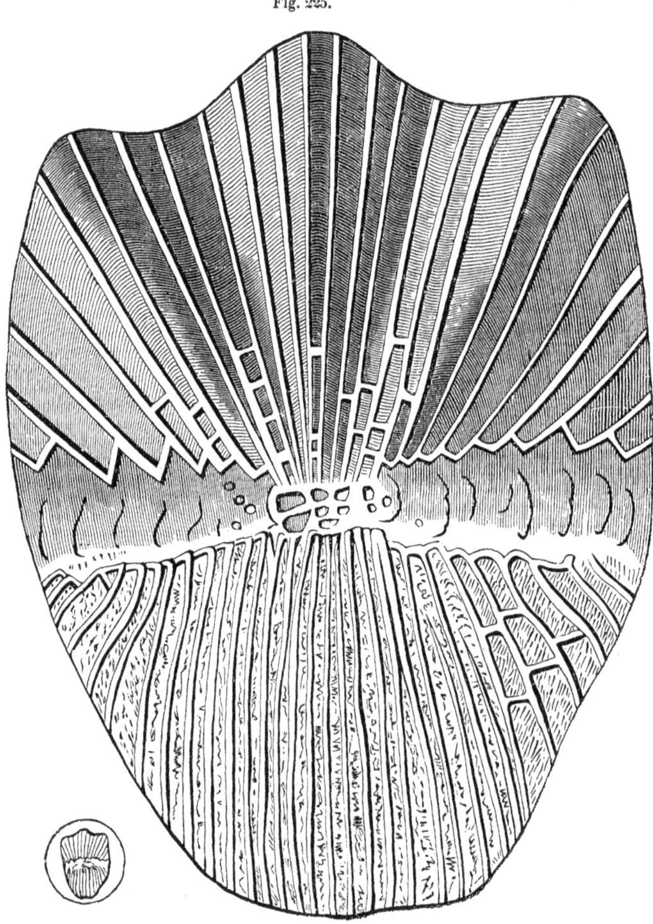

Fig 224.

parrot-fish of its natural size ; and in figure 225 the same is shown, as it appears when considerably magnified. The position of each layer is here indicated by the numerous waving lines that cover the surface of the scale, and the ribs and flutings which branch out from the middle portions, are very strongly marked.

Fig. 226.

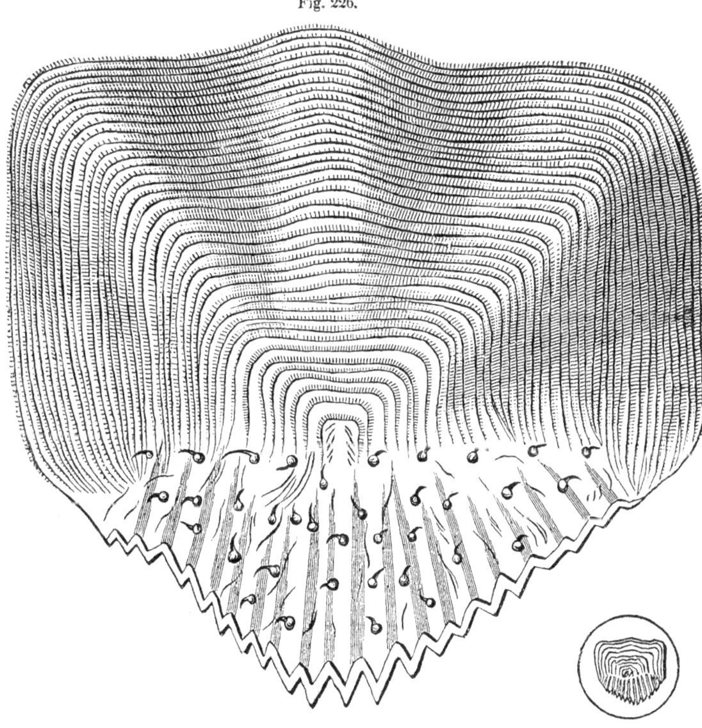

Fig. 227.

Figures 226 and 227, represent the scale of the *sea-perch*, both magnified and of its natural size. This scale is broader in proportion to its length than that of the parrot-fish, and unlike the latter, is destitute of the radial divisions. The edges also of the component strata, as seen in the magnified figure, are not bounded by curved lines, but are serrated, presenting an appearance like the teeth of a saw. The lower part of the scale is likewise notched along the edges, which gradually approach each other, and unite at the base.

A scale of the haddock is delineated in figures 228 and 229, in the first of which it appears of its natural size, and the other displays a magnified view of the same.

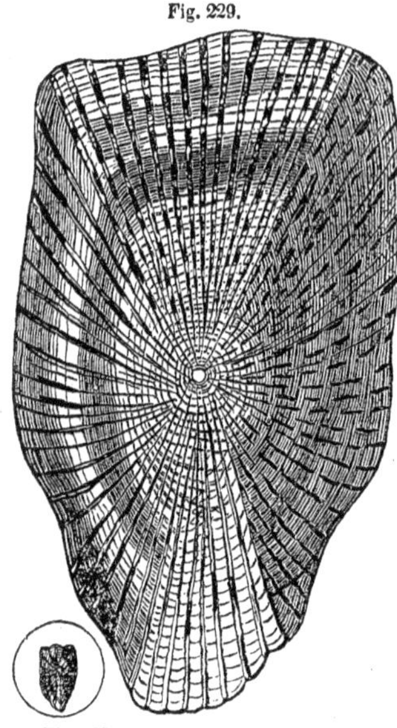
Fig. 229.

Fig. 228.

It is a beautiful scale, resembling a shield in form, and the entire surface is covered with numerous radial lines crossing the concentric strata.

A scale of the roach is exhibited of its natural size in figure 230, and the same magnified is shown in figure 231. It is seen to differ from all the preceding scales in many particulars. The broad flutings rise fan-shaped from the centre of the scale, flanked on either side by numerous concentric lines, which indicate the position of the edges of the overlapping strata. These lines do not terminate at the extreme radial branches, but pass across them as in the case of the parrot-fish. The lower portion of the figure, which is apparently covered with teeth, represents the root of the scale, by which the latter is attached to the fish. The distance, *a*, *b*, between two of the radial lines, as measured by the micrometer, is *one-fiftieth part of an inch.*

In figure 232 and 233, another scale is shown, both in its real and magnified dimensions. It is the scale of the flounder, and resembles in some points that of the roach. The series of concentric lines on each side, crossing the radial divisions, are mainly alike, and the fan-like flutings are seen in both, only their divergence is far greater in the scale of the roach than in that of the flounder. Owing to this circumstance the forms of the scales are different, that of the roach being broader than it is long—while in the case of the flounder, the length exceeds the breadth. The distance between two of the radical lines, *a* and *b*, in figure 233, is the *one-three hundred and tenth part of an inch.*

Figure 234 is a magnified portion of the skin of a sole-fish, viewed by reflected light—the dark ground representing the skin, and the lighter parts the upper protruding portions of the scales, which exhibit a beautifully serrated appearance. Figure 235 is the same, of its natural size.

THE INTERNAL ORGANS OF RESPIRATION OF THE SILK-WORM.—The mode in which insects respire, is very different from that which exists among the higher orders of animals. They are not possessed of lungs, neither do they breathe through the mouth; but inhale the air through numerous orifices, called *spiracles*, with which are connected respiratory tubes, that extend in minute ramifications to every part of the body. These tubes, which are divided into two classes, consist of three coatings. The first, or external envelope, is a membrane comparatively thick, strengthened by a great number of fibres, which form around it numer-

Fig. 231.

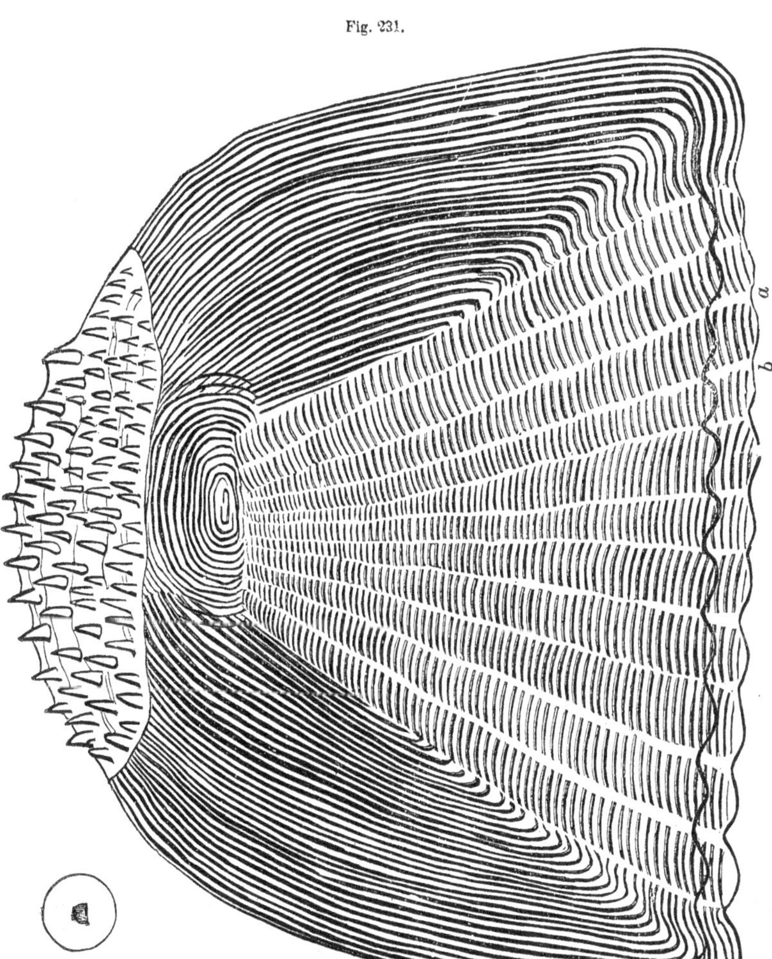

Fig. 230.

ous irregular circles. The second tunic is a tissue more delicate and transparent, and the third is formed of a cartilaginous thread, wound spirally upon itself. In figure 236, are exhibited portions of the respiratory tubes of the silk-worm, considerably magnified. The fibrous structure is at once perceived, and the end

Fig. 233.

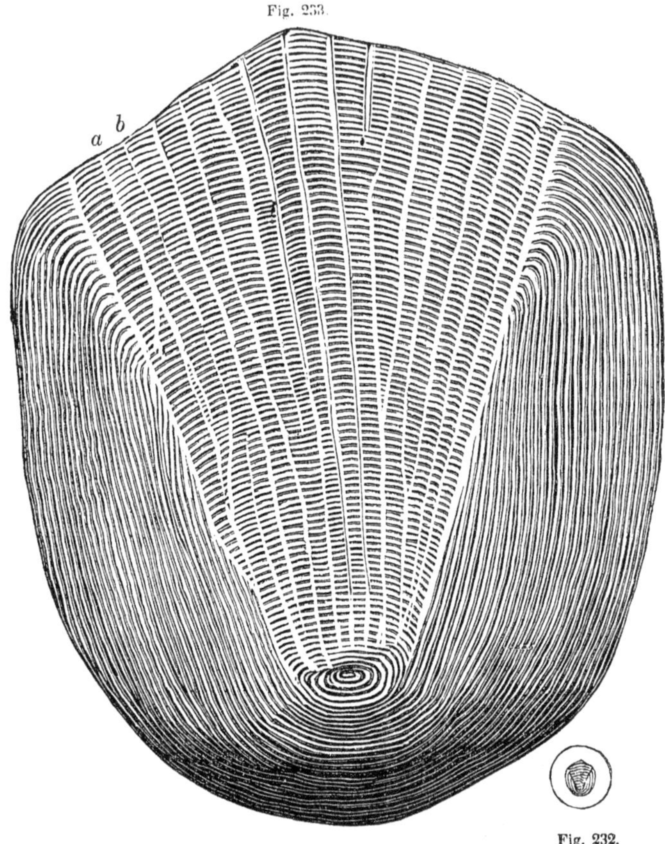

Fig. 232.

of a spiral thread is seen at *a*, the thickness of which measures only *one-fifteen thousandth part of an inch.* The breathing tubes, with the branches into which they subdivide, are very numerous in some insects, amounting to more than eighteen hundred, and the ramifications become at length so exquisitely fine, that the most powerful lenses fail to detect them.

Magnified Flea.—This wonderfully active little creature is a good object under a moderately magnifying power. The head is small, and covered with a shelly plate, and on either side gleams a brilliantly dark eye, the pupil of which is encircled with an iris of a greenish hue. Behind the eyes are two small

Fig. 235.

Fig. 234.

cavities fringed with hairs, and which are supposed to be the ears. Below the head are three jointed members, which are the feelers and piercer of the insect. The piercer consists of a tube and tongue, and on each side of the latter is a sharp lancet-like blade, with which the flea punctures the skin of its victim.

All the legs are fringed with hair, and are terminated by claws. In order to leap, the flea folds up its six legs, and then instantaneously extending them, makes its spring, exerting its whole strength at one effort. The body of the insect is encased in an envelope consisting of overlapping plates, symmetrical in form and arrangement. Along the back and under the belly, the plates are studded with hairs, equally distant from each other, and ranged in a line along the middle of the plate. The distance between two contiguous hairs in the same row, is about *one-five hundredth part of an inch.* The plates near the head are likewise fringed with hairs.

Fig. 236.

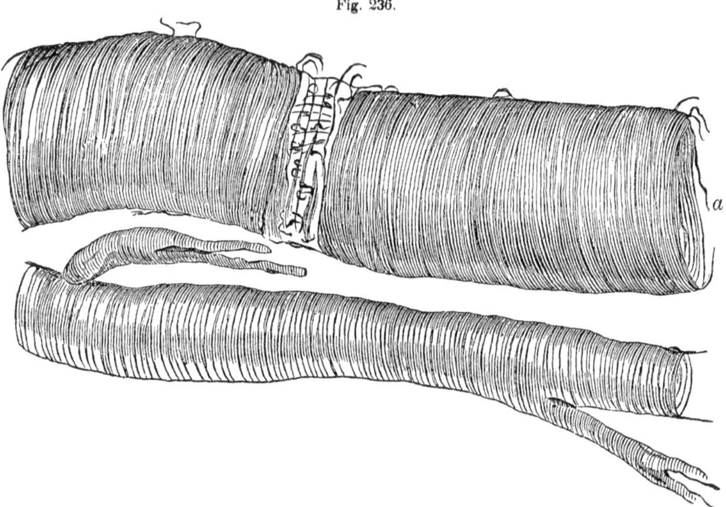

The strength of the flea is very great; for at the fair of Charlton in Kent, in the year 1830, a man exhibited three fleas harnessed to a carriage fifty times their own bulk, which they pulled along with great ease; another pair drew a carriage, and a single flea a brass cannon.

A Mite Magnified.—Upon carefully viewing with the naked eye the fine dust of figs, or decayed portions of old cheese, round, living specks will frequently be seen, moving slowly and with difficulty among the atoms by which they are surrounded. These specks have received the name of mites, and are so small that they easily elude observation. When magnified under the solar microscope, their images are seen moving around upon the screen, endeavoring to avoid the glare of the light. They then appear of considerable dimensions, and their several members and parts are distinctly revealed. A magnified mite is delineated in figure 238. Each of its numerous legs are seen to consist of several joints; its body is oval, tapering towards the head, which is furnished with antennæ, and its surface is covered with numerous long and slender hairs. It is naturally a disgusting creature, and the unpleasant associations connected with it render it still more so.

Globules of Blood.—When a drop of flowing blood is taken from the veins of an animal and spread over a glass slide, it is seen to consist of a *fluid*, together with numerous *rounded particles* termed *globules*. These globules enable the observer to detect the motion of the blood, to establish the fact of its circula-

Fig. 238.

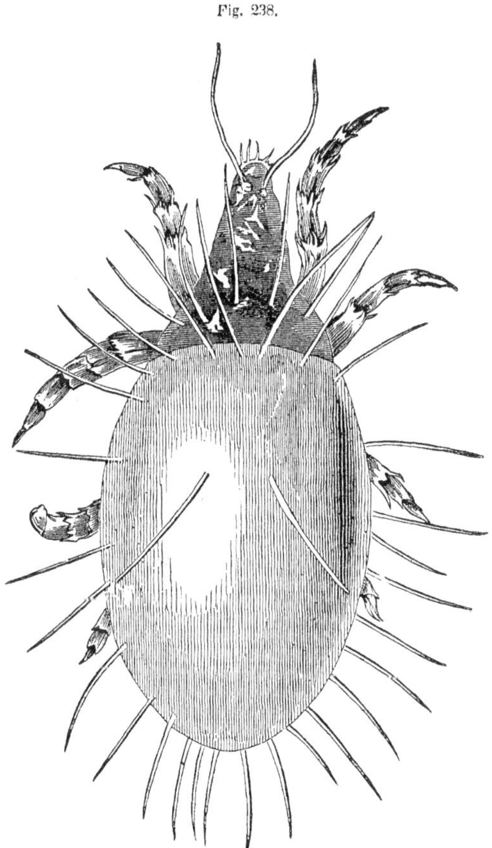

tion, and to mark its course as it speeds along through the arteries and veins. The globules are very abundant in the blood, each drop being filled with many thousands, and yet, small as they are, they have been accurately studied and examined. They are divided into three kinds, the red and white globules, and other smaller atoms, which have received the name of *molecules*. The red globules far excel the white in number, and appear, as they roll through the centre of the blood-vessels, to constitute the greater portion of the fluid.

In man and in most Mammalia, these atoms possess a round, flattened form, like that of a coin, with a slight depression towards the centre. The position of this depression is indicated by a dark spot, and its depth depends upon the magnitude of the globule. In figure 239, the red disks of the human blood are delineated as they are revealed when subjected to a high magnifying power. They are here seen promiscuously scattered over the surface, though they are often beheld united together by their flat surfaces, and forming little bead-like rows of crimson atoms. The central depression is distinctly visible in the several atoms. The size of the red globules is subject to much variation, even in the same animal. In human blood it ranges, according to the best authorities, from *one-thirty-five hundredth part of an inch* to *one-forty-five hundredth*, though an eminent observer has found their average diameter to be as great as *one-twenty-eight hundredth of an inch.* Their size in the elephant is about *one-twenty-seven hundredth of an inch*, and in the napu-musk deer only *one-twelve thousandth.*

Fig. 239.

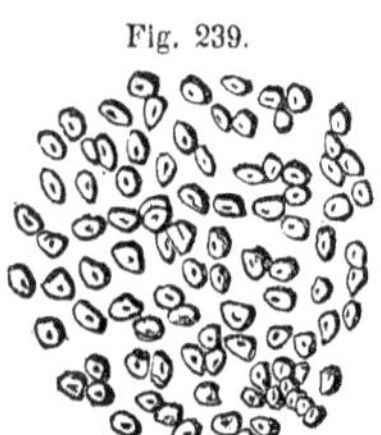

The blood-disks in birds, reptiles, and fishes, differ from those of Mammalia, in being larger, and their shape is also *oval* instead of *round;* moreover, in place of being depressed at the middle, they swell out on either side, owing to the fact that the centre of the atom is composed of matter more solid than the other portions. In birds, the length of the oval disk varies from *one-seventeen hundredth of an inch* to *one-twenty-four hundredth*, and the breadth from *one-three thousandth of an inch* to *one-forty-eight thousandth.* In the case of frogs, the longer diameter is about *one-thousandth of an inch* in extent, while in fishes these globules are for the most part larger than those of the frog. The white globules in man and the Mammalia, are usually larger than the red, but like the latter, they differ in magnitude. Their average size, when examined in the blood, is estimated at about *one-twenty-six hundredth part of an inch.* In the blood of reptiles, and in that of the frog in particular, the relation that exists as to size between the red and white globules, is reversed; the latter being in these cases two or three times smaller than the former. These two classes of atoms differ also in respect to form, since the white blood particle is always globular throughout the whole animal kingdom: a nucleus, consisting of matter more solid than the rest, is also found in the white globule, instead of a central depression as detected in the red.

The third class of atoms, termed molecules, have been regarded as the elements out of which the other two kinds are formed. They are found in great quantities amid the blood, existing singly, and also in small masses of an irregular form. Their minuteness far surpasses that of the other atoms, since they scarcely ever exceed in diameter *one-thirty thousandth part of an inch.*

THE WEB OF THE FROG'S FOOT.—When the web of a frog's foot is examined with a high magnifying power, it exhibits a beautifully tesselated ground, intersected by blood-vessels and minute capillaries, that wander over its surface. In these the circulation of the blood is distinctly seen, the fluid coursing swiftly through the arteries, but moving with less velocity through the veins. The red globules appear in immense numbers, and in the minute ramifications of the vessels, are seen rolling along in single rows, with here and there a white globule scattered among them, not more than half as large as the red oval disks.

The velocity of the blood is not uniform; for the current is observed to be subject to sudden momentary checks, after which it again flows on with its former speed. In figure 240, is delineated a portion of the web of a frog's foot,

Fig. 240.

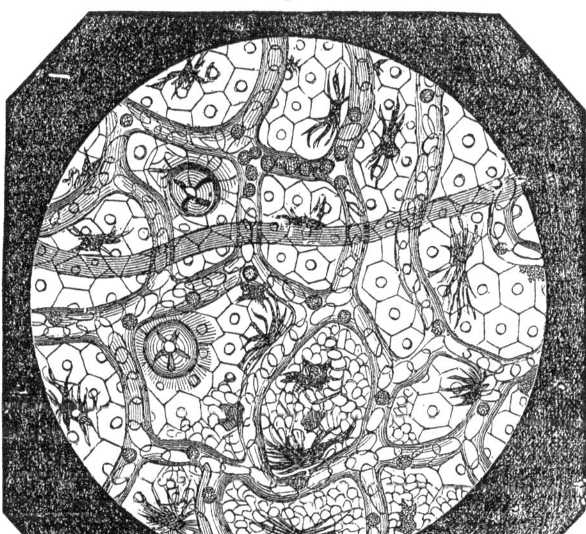

magnified *three hundred and fifty diameters.* The web has the appearance of mosaic work, being divided into beautiful hexagonal figures with a nucleus in the centre of each. The most minute ramifications of the blood-vessels are here seen standing prominently forth, and within them the blood globules are clearly revealed—the large oval disks representing the red atoms, and the small round ones the colorless particles. An idea may be gained of the size of the capillaries by recollecting the length of the red globules in the blood of the frog.

POLLEN.—The pollen of flowers which appears as a fine dust to the unassisted eye, is shown by the microscope to be an assemblage of organized bodies, possessing regular figures, and varying in size, form, and color, according as they are taken from different plants. The color of the pollen is usually yellow; but it is frequently found to be of a purple, white, blue, and brown hue; and in some flowers it appears in the form of clear, transparent grains. The surface of the particles in some instances is smooth, and in others rough, and in many cases it is studded with delicate spines or thorns. The pollen is contained in a receptacle termed the anther, which at the proper time opens and liberates the imprisoned particles. These are not unfrequently borne upon the atmosphere to a great distance; for trees have been known to be fructified by pollen, which must have been wafted through the space of three miles. The number of particles contained in each anther, varies from a few hundreds to several thousands.

When the grains of pollen are viewed with a microscope, at the time they are fully matured, they are seen to separate, and an oily liquid flows from the interior. A similar result occurs if a grain of pollen is thrown upon the surface of water. It there gradually swells and at last bursts, when a liquid escapes from the atom, which spreads in a thin film over the surface of the water in the same manner as a drop of oil. This liquid has been regarded as the fructifying matter of the plant. An anther of the mallow is delineated in figure 241, and the grains of pollen that it bears are indicated by the round spots in the middle of the drawing. Figure 242, shows the atoms of pollen more highly magnified.

Fig. 241.

Fig. 242.

The pollen of the morning-glory is delineated in figure 243. It appears under the microscope of a spherical form, like a small pea, with the surface thickly set with minute spines. It is of a pearly white color, and appears to be composed of an assemblage of small cells, the partitions of which are indicated by the light which passes through them, on account of their transparency; and in the figure their situation and mode of arrangement are distinctly marked by the lighter parts of the drawing. The real diameter of these particles of pollen is the *one hundred and twenty-fifth part of an inch.*

Fig. 243.

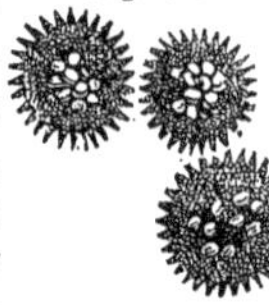

INDIAN CORN.—The pollen of the Indian corn is exhibited in figure 244. In shape, the grains resemble those of buckwheat; the central parts are thin and transparent, and are probably cells filled with fluid. The length of a side of one of these atoms does not exceed the *eight hundredth and thirtieth part of an inch,* and the diameter of the small central cell, is less than the *three thousandth part of an inch.*

Fig. 244.

FUSCHIA.—In figure 245, several particles of the pollen of the fuschia are displayed, magnified one hundred and ten times. They are of a brown color, and are similar in shape to the pollen of corn when viewed, as they are diffusely spread in their natural state over the surface of a slip of glass. But when they are immersed in a layer of balsam between two plates of glass, they assume a different form, and little round appendages are then distinctly discerned like handles, at each corner, as exhibited in the figure. One of the largest of these specimens measures in its longest extent, *the three hundred and sixtieth part of an inch.*

Fig. 245.

Fig. 246.

SWEET PEA.—The pollen of the sweet pea is delineated in cut 246, as it is revealed under a considerable magnifying power. It appears as a collection of brown oval grains, with central cells of the same shape, placed lengthwise of the grains, and their positions are indicated by the light lines in the several figures: the clear fluid with which the cells are filled, rendering them transparent. The length of one of these atoms is the *five hundredth part of an inch*, and the breadth *one-six hundred and twenty fifth part.*

FERN SEED.—The seed of the fern affords an interesting object for the microscope, and in cut 247 a sketch of various parts of the plant is presented, which is taken from Swammerdam. *a* represents a stalk of fern, the leaflets of which at the lower part of the stem are thickly covered upon the back with the seed-vessels of the plant. At *b* and *c*, two of these seed-vessels are seen highly magnified. The stalk of the seed-vessel is smooth, but where it unites with the pods it changes into a strong cross-ribbed thread, which completely encircles the pod and holds it firmly together. This singular cord is shown at *b*, as it appears edgewise, and in *c*, a side view is presented, with the enclosed pod, the shaded line across the latter indicating the position of a natural fissure in the pod. When the seed is ripe, the circular elastic cord is straightened out, and in the process of unbending, opens the seed-vessel, completely separating it into two parts through the natural fissure, and forming two hemispherical cups, which are attached by short stems to the elastic cord. This stage of development is seen at *d*, where the straightened cord and the two open hemispherical cups are delineated. By imagining the elastic cord to be bent back into its original shape, it is evident that the edges of the two cups would unite, and that the figure *d* would re-assume its original form as shown at *c*. At *e* a seed-vessel is shown, in which an opening has been made, and a portion of the enclosing membrane

Fig. 247.

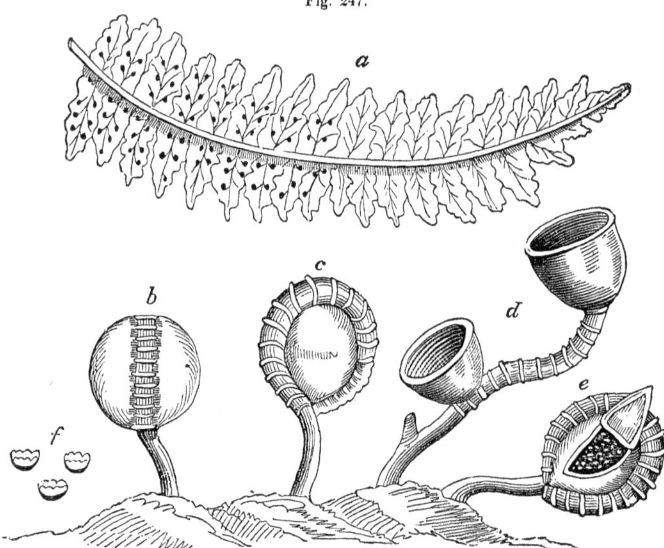

thrown back, in order to exhibit the seeds grouped in their natural position within the pod. Three seeds, out of forty, taken from the same pod, are represented at *f*, very highly magnified.

www.ingramcontent.com/pod-product-compliance
Lightning Source LLC
LaVergne TN
LVHW021359110826
845150LV00007B/1715

* 9 7 8 1 4 2 5 5 1 3 2 3 8 *